the world of caves

A C Waltham

G.P. PUTNAM'S SONS · NEW YORK

© Orbis Publishing Limited, London 1976
SBN: 399-11733-4
Library of Congress Catalog Card Number: 75-37228
Printed in Italy by IGDA, Novara

*Endpapers: Easter Grotto of Easegill Caverns in the
English Pennines*

The photographs in this book are the author's, with the
exception of those listed below:
Page 6 *Foto V. Radnicky;* **10** (top left) *H. Fairlie-Cuninghame;* **16** *H. Fairlie-Cuninghame;* **17** (bottom) *C. D. West-
lake;* **19** (left) *Spectrum Colour Library;* **20** *J. R. Wooldridge;*
21 (top) *J. R. Wooldridge;* **22** *Photo Loic-Jahan;* **26** *H. Fairlie-
Cuninghame;* **27** *A. J. Pavey;* **28** *J. R. Wooldridge;* **29** *H.
Fairlie-Cuninghame;* **31** *Bruce Coleman Limited/R. D. Ellis;*
32 (bottom) *H. Fairlie-Cuninghame;* **33** (bottom) *H. Fairlie-
Cuninghame;* **34** *Ardea Photographics;* **36** *Bruce Coleman
Limited/L. R. Lawson;* **37** *Bruce Coleman Limited/Alan
Blank;* **38** (top) *Bruce Coleman Limited/C. Bisserot;*
(bottom left) *Ardea Photographics;* (bottom right) *Aquila
Photographics;* **39** *Foto V. Radnicky;* **42** (bottom) *W. R.
Halliday;* **43** (top) *Ardea Photographics;* (bottom left) *J. M.
James;* (right) *Bruce Coleman Limited/Jane Burton;* **44** *R.
Everts/Pictor;* **46** *Ardea Photographics;* **47** (top) *Günter Heil/
Pictor;* (bottom) *Ardea Photographics;* **48** *Spectrum Colour
Library;* **49** *E. Hummel/Pictor;* **50** *Ardea Photographics;*
51 (top) *I.G.D.A.;* (bottom) *J. Allan Cash Limited;* **52** *C. M.
Dixon;* **53** *French Government Tourist Office;* **54** (left)
Spectrum Colour Library; (right) *Bruce Coleman Limited;*
55 *Colorific;* **56** *J. Allan Cash Limited;* **57** *I.G.D.A.;*
58 *Ardea Photographics;* **60** *Douglas Dickins;* **61** *Konrad
Helbig/Pictor;* **66** *I.G.D.A.;* **67** *Photo Loic-Jahan;* **69** (top)
Picturepoint Limited; **71** (bottom) *R. S. Harmon;* **72-3**
Moshe Zur/Pictor; **76** *Spectrum Colour Library;* **77** (top)
Pictor; (bottom) *S. Talent;* **80** (left) *P. L. Smart;* **81** (top)
W. R. Halliday; **82** (top) *R. S. Harmon;* (bottom) *W. R.
Halliday;* **83** (top) *W. R. Halliday;* **84** (left) *R. Sheridan;*
93 (top) *A. G. Latham;* (bottom) *R. S. Harmon;* **94** *J. R.
Wooldridge;* **97** *Courtesy of the Italian State Tourist Office,
E.N.I.T.;* **102** (bottom) *M. Serban/Zefa;* **103** *Foto V.
Radnicky;* **104-5** *Akiyoshi-do Museum;* **106** *Ardea Photo-
graphics;* **107** *J. C. Whalley;* **108** *H. M. Beck;* **109** (top)
L. N. Robinson; (bottom) *A. J. Pavey;* **110** *C. D. Westlake;*
114 (left) *C. D. Westlake;* **115** *J. C. Whalley;* **118** (bottom)
D. Brook; **122-3** *J. Russom*

The diagrams were compiled from original surveys by the
following: *United States National Park Service* (**83**);
Association of Mexican Cave Studies (**84**); *Commisione Grotte
'E. Boegan'* (**97**); *Groupe Spéléologique Florentin* (**98**);
Postojna Krasa Institut (**99**); *University of Leeds Speleo-
logical Association* (**113**); *McMaster University Caving and
Climbing Club* (**119**)

the world of caves

Contents

The Subterranean World

Matavun is a quiet hamlet in northern Yugoslavia. The white walls and red roofs of the small cluster of houses are barely noticed in the open countryside. Village life in this region of mainly farming communities is peaceful. But all through the summer, groups of tourists gather in the main street and periodically walk off over the hill to the west. The tourists have come to Matavun because almost directly beneath the houses lies the cave system of Skocjanske Jama, one of the wonders of the world.

The road out of the village leaves the houses behind, and soon a path branches off to the left and descends a steep wooded slope. Through the trees it can be seen that the descent leads into a huge, rounded, conical depression. There is no way out at the bottom, except through a steel door into a mined tunnel. A string of light bulbs beckons into the darkness, and after a short distance the confines of the artificial tunnel give way to the spacious vault of the natural cave. This first view of Skocjan Cave is of a wide gallery with a dry sand floor, descending gently; white and pale orange stalactites hang from the ceiling, and rows of massive stalagmites line the path. Round the corner the passage is larger—10 metres (30 feet) high and a little wider, its gently arched roof supported by the massive, tiered stalagmite columns which have joined the stalactites hanging from above. Like an ancient street flanked by buildings of an extravagant bygone architecture, the path snakes between the ranks of stalagmites, and each turn reveals a new composition of these cave decorations. Finally a straight stretch of the passage reveals rows of stalagmites, which end in a black space.

A distant rumble of cascading water breaks the silence of the caves. The path leads right to the edge of the blackness and a low steel fence guards a precipitous drop. Nothing whatever can be seen in any direction—then the lights are switched on to reveal the true splendour of the Skocjan Cave. An immense cave passage becomes visible: it is 60 metres (200 feet) down to the river; it is another 30 metres (100 feet) up to the roof and the same distance to the far wall. The path and the end of the stalagmite passage are perched in a hole more than half-way up the near-vertical wall of this vast river cave; and this lends a tremendous impact to the first sighting of an almost unbelievably large natural tunnel in the limestone.

But the path continues—not down to the floor of the river cave, but along a ledge cut out of the wall. It offers magnificent views down to the river, until it turns out over a narrow bridge, which leads across to the far wall of the cave. From it can be seen the cascades and pools of the river far below, the great floodlights down there almost struggling to light the vastness of this subterranean world. Downstream the river disappears into darkness, but upstream it is lit, for that is the way the path goes—clinging to another series of terraces and ledges. Farther on the walls of the cave are even more broken—by black openings and galleries which lead off; the limestone begins to resemble a mammoth rabbit warren. The path does not stay in the river gallery but loops round some of the side passages, through more stalagmite-decorated chambers.

Then from one viewpoint the glimmer of daylight can be seen reflected from the glistening wet walls of the river cave—the first daylight since the entrance over a kilometre (nearly a mile) behind. The path follows its own passage round to the outside world, and there it again breaks out onto a ledge part-way up an enormous open space. But this one is lit by daylight, for it is a huge crater-like depression. Far below is the river crossing its floor, and white cliffs of limestone tower 150 metres (500 feet) above to the woodlands and fields around Matavun. The ring of

Left: Stalactites decorate a roof in the huge Punkevni Cave system in Czechoslovakia

7

Above: The River Reka pours through one of a series of natural arches before plunging into the enormous entrance of the Skocjan Cave in northern Yugoslavia
Right: A steel bridge carries the tourist path across the main river passage deep inside Skocjan. Far below, the Reka cascades into the darkness of a massive rift cave
Far right: Blocks of bare limestone furrowed by rainwater lie on the surface of the Astraka Plateau in northern Greece

cliffs is broken on one side by a jagged gash where the river pours out of a great cave arch—but its daylight journey is brief for, on the opposite side of the depression, the river hurtles back underground into the entrance of the Skocjan Cave itself. The path winds its way up a flight of steps up the gentlest slope; the river and its own special world of darkness are left behind. On the few minutes walk back to the houses of Matavun the visitors are left with a feeling of amazement— amazement not only that caves can exist on such a grand scale, but also that there is so much variation and contrast in the underground world of limestone.

Like almost every other cave in the world, Skocjanske Jama and everything about it—from the massive canyon passage to the forests of stalagmites—was formed just by running water.

When rainwater hits the ground it may do a number of things: some collects in streams and

rivers and flows away; some evaporates back into the atmosphere and some is utilized by plants; and some sinks into the ground. This groundwater, as it is then known, seeps through the soil and down fissures in the rock, and eventually returns to the surface through small springs or invisible seepages into river beds. Most of this groundwater moves so slowly that it cannot erode or change the rock as it passes through—unless the rock is limestone. For limestone (and a few other less common rocks) is soluble in water.

Not surprisingly, the processes of limestone solution and cave formation are many, varied and complex. Some idea of their nature may be gained by tracing the path of just one drop of water as it passes through a hill of limestone where cave formation has reached a fully mature stage, where new caves are still forming, even though older caves do exist and in some places are already being filled in.

The drop of rainwater falls from the sky in a fairly pure state, and such pure water is only capable of dissolving a tiny amount of calcium carbonate—the chemical which forms the mineral calcite, which in turn makes up the rock limestone. It is therefore very significant that our raindrop lands on soil and then slowly seeps down through the surface layers of broken rock and decayed vegetable matter. As it does so it absorbs a considerable amount of the gas carbon dioxide from the 'soil air'—and it is this that gives the water the major part of its power to dissolve limestone. The amount of calcium carbonate that can be dissolved in water increases with the amount of carbon dioxide dissolved in the water.

This particular drop of water then seeps through the soil and enters the underlying bedrock through a tiny fissure in the limestone. The next part of its journey is very slow—probably only half a metre (a foot) a day as it passes through a network of narrow openings. But as it does so, highly charged with carbon dioxide from the soil, it dissolves the limestone until it is chemically saturated with calcium carbonate. It has made its first contribution to cave development on a minute scale. Eventually this network of micro-fissures ends in the roof of a pre-existing open cave passage, and the drop of water finds itself hanging from the roof, surrounded by the atmosphere of the cave.

Chemically the cave air is very like normal air—with a very low carbon dioxide content. But the drop of water gained a far greater share of carbon dioxide from its passage through the soil. It is therefore out of equilibrium, and the result is diffusion of carbon dioxide from the water into the cave air. Then the calcium carbonate content of the water is left out of balance with the carbon dioxide—so it too comes out of the water, as a precipitate of calcite. Hanging on the cave ceiling the drop of water therefore contributes to the growth of a stalactite. Before equilibrium is reached more water seeps down through the limestone, and the drop is pushed off the cave roof. It falls to the floor and deposits its still excessive load of calcite, building up a stalagmite.

Now bearing a much smaller content of dissolved mineral, the water runs off over the cave floor and joins an underground stream. Cascading down the cave passages, through pools and over waterfalls, the water now erodes the limestone in a new way—mechanically. Just like a river above ground, this turbulent water scours out its channel, chipping away bits of rock, while at the same time it still carries away a large amount of limestone in solution. Its rapids and waterfalls cut a deep canyon cave through the rock, but lower down the stream enters a flooded zone. There it flows through a completely submerged passage. Eventually it has to flow uphill to leave this massive U-tube in the cave, and only the pressure of water behind keeps it going. Its

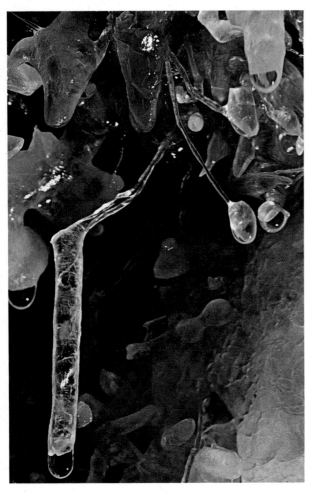

movement is sluggish and it has lost the power to abrade or scour the cave walls, and it is also saturated with calcium carbonate.

Even though it is chemically saturated, this stream is still capable of dissolving further limestone. On its route through the permanently flooded caves it gains a tributary stream at a submerged junction. The water of this stream is also saturated with calcium carbonate, but it has had a different journey through the limestone, so it is chemically different from the water of the main stream. The detailed nature of the solution of calcium carbonate and its non-proportional dependence upon the carbon dioxide present means that the mixing of these two saturated waters gives a water of new chemistry which is unsaturated. More limestone can therefore be dissolved and the water is yet again capable of cave development.

Eventually the stream flows out of the cave at a spring, and the surface river carries away its dissolved load of limestone, ultimately to the sea. One drop of water on its route from daylight has therefore dissolved some limestone, reprecipitated some of it, abraded some more and dissolved yet more. With such a range of processes active in limestone, it is hardly surprising that caves show such amazing variety of size, shape, pattern and decoration.

The first contrast that strikes the visitor to almost any cave—Skocjan Cave for example—is between the silence of the stalactite-draped dry tunnels and the noise and rumble of the stream or river passages. The dry caves seem to be dead and timeless, compared to the life and activity of the water caves. In fact very few cave passages are completely dead—nearly all are changing in some way and to some extent—but it certainly is true that it is the river caves where the most spectacular erosion is taking place.

Just like a river in daylight, a cave river carves and erodes its bed out of the rock. Cascades, waterfalls and rapids cause powerful currents to scour the bedrock. Sand, gravel and boulders are moved along by the water, giving teeth to its erosive bite. On bends and in cascade pools the streams hurl the debris at the rock walls, carving them away. A cave river does all these and in addition it attacks the limestone chemically, carrying it away in solution. Most of this erosive effort is concentrated on the floor of the stream channel, so a stream or river tends to cut downwards and form a canyon. Caves of this type are known as canyon passages; the river cave in Skocjan is a fine example. Cave scientists refer to them as vadose caves or vadose canyons, vadose meaning that the cave has been cut by free flowing water.

When a river out in daylight cuts its bed downwards it tends to form a canyon, except that the walls are broken down by further erosion. Weathering of the rock, rainwash, and the cutting

power of tiny rivulets wear back the steep walls so that a V-shaped valley is formed where broken rock is fed downwards and carried away by the river. The world's genuine canyons are generally formed where powerful rivers are cutting downwards but there is almost no water erosion and destruction of the walls. Hence the Grand Canyon in the United States is formed where the Colorado River crosses, and cuts a nick through, an almost waterless and erosion-free desert. Similarly the immense canyon passage inside Skocjan Cave is formed because it has a roof on it—so protecting its walls from weathering and rainwash.

In steeply descending cave systems, the canyon passages are broken by waterfall shafts where the stream cuts down to lower levels in the rock. The greater speed of the falling water, its hammering effect in big waterfalls, and the corrosive action of the swirling spray which is such a feature in cave waterfalls account for the shafts tending to be much larger than the canyon caves which connect with them. Some of the narrow little canyon passages, cut by small streams, are so tight that a caver has to squeeze his way through. Many of these then lead to roomy shafts where the caver swings free on his ladder as he climbs down through the spray without touching the walls and then has to squeeze into another tiny canyon which is the only way out at the foot of the

Left : Inside the Noisy Water Cave in central Jamaica, Cave River carves its way through the limestone
Below : Beneath its stalactite-draped roof, the water in the Golding River Cave, also in central Jamaica, barely moves—but nevertheless continues to erode the limestone by chemical solution

shaft. Some caves have a whole series of shafts, so that the passages descend through the limestone like giant staircases; the Cambou de Liard cave system in the French Pyrenees has more than 50 waterfall shafts, descending to a total depth of just over 900 metres (just under 3000 feet). None of the shafts in the Cambou de Liard is particularly deep, but such chasms can be of impressive proportions. Deep in the Hochleckenhöhle Cave of Western Austria there is a waterfall shaft 370 metres (1200 feet) deep.

Vadose canyons and waterfall shafts are where caves can be seen to be formed—yet they provide only half the picture. The limit of exploration in so many caves is where the water meets the cave roof. Beyond there the cave passages are permanently flooded and completely full of water. Only cave divers ever see these passages, and their air supplies limit the extent of their exploration, even though some cave rivers flow through many kilometres (or miles) of flooded passage. This largely unseen section of the world of caves is again the scene of erosion and cave development, though of a very different style from that of the canyon caves.

With the water completely filling these flooded caves, the flow rates tend to be much lower than in the descending canyon passages, so little sediment can be transported and relatively little abrasion of the cave walls takes place. Instead

solution becomes more important; hence the significance of mixed-water-corrosion. This is the mechanism whereby cave streams meeting in this flooded zone are given boosts to their corrosive power; even when two streams, each saturated with calcium carbonate in solution, meet and mix, the combined stream is capable of yet more solution of the limestone. As the water moves slowly through the flooded caves, completely filling the passages and being driven along by hydrostatic pressure, it does not just carve away its floor—instead it erodes floor, walls and roof equally, by solution and perhaps abrasion by very fine, suspended sediment. So instead of cutting a canyon into the rock, the stream dissolves out a cylindrical tube, a tube being dynamically the most efficient shape for a completely filled water passage. Caves of this type are referred to as phreatic tubes by cave scientists, phreatic meaning that they have been cut by water flowing under pressure—in complete contrast to the vadose caves cut by free flowing streams with an air surface above the water.

As the water in phreatic caves flows under pressure, it can flow uphill. Cave divers exploring upstream in the flooded zone of Gavel Pot in the English Pennines have found the water flowing up a vertical cylindrical shaft more than 40 metres (130 feet) deep. The flooded zones of cave systems can only exist because the water has to

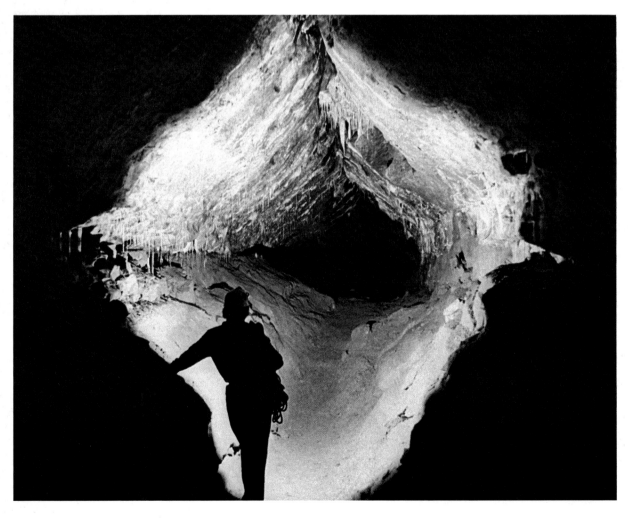

Left: The Minaret Passage in Lancaster Hole, England, resembles the shape of a diamond. The cave was formed by water working its way along the fractures still visible today

flow uphill where it cannot freely drain out of the limestone. Cave passages may be developed uphill because of local geological variations in the limestone—giving local phreatic zones. But more commonly phreatic caves are formed where water loops down through the lower levels of a limestone block and then has to flow uphill to get back out onto the surface. Immense flooded U-tube systems are, therefore, formed with the outlet springs providing the level up to which the cave passages are kept full of water. Hence so many cave systems have vadose canyon passages in their upper sections and phreatic tubes in their lower reaches. The great cave of Padirac in Central France has a magnificent great canyon passage many kilometres (or miles) long. But at its downstream end there is a sump—where the water meets the cave roof; and beyond there lie more kilometres (or miles) of flooded phreatic tube, all the way to the St George's Spring where the water flows out over the lip of a flooded shaft and meets daylight, just before joining the River Dordogne.

Tubular cave passages are easily recognizable as phreatic in origin. The active ones flooded to the roof are seen by few, yet dry phreatic tube passages are common throughout the world. The high level stalagmite cave on the tourist route into the Skocjan Cave is one. The great rounded tunnels of the Eisriesenwelt Caves of Austria, and

Right : Hensler's Stream Passage in England's Gaping Gill Cave system is a fine example of a vadose canyon passage. The stream which initially flowed through the limestone at the level of the flat roof has subsequently cut its way down to the present floor level
Below : The rounded profile of the main passage in England's Peak Cavern indicates that it was formed when completely filled by water flowing under pressure. It is a phreatic cave—even though it now only carries a tiny free-flowing stream

Above : Holes-in-the-Floor Passage in Castleguard Cave, Canada, is formed by a beautifully rounded phreatic tube with a narrow vadose canyon cut in its floor. The caver is standing on a natural bridge of dried mud—a remnant of a layer of mud that once half-filled the passage Opposite : The daylight shaft of Alum Pot in the English Pennines is formed where the water cascades down a massive fracture in the limestone

ment of the valley. Gargantua is older than the valley. It was formed long ago when the whole limestone mountain was flooded with water and draining out at a high level. Then glaciers cut the valley, the limestone drained out into the new low level rivers, and the cave was literally left high and dry. The entrances now lie where the valley wall has truncated the cave passages, their original continuations having been removed in the limestone gouged away by the glacier.

Gargantua and the many other phreatic cave systems, such as Mammoth and Peak, have therefore been abandoned by their water—they are in fact fossil remains of long-gone drainage networks. Clearly all the phreatic caves now dry and visible have been abandoned, but in the same way vadose canyon caves can be abandoned by their streams. Water flowing through fissured limestone, in vadose cave systems, continually tends to leak down into lower levels of fractures and fissures, thereby developing new cave passages at the expense of those abandoned at higher levels. Cave rivers differ from surface rivers in that with each new stage of erosion, instead of the older features being themselves removed, they are left preserved in the limestone. External changes, such as valley cutting, and internal changes such as the opening up of new fissures may all result in the drying out and fossilizing of cave passages, and these dry passages in turn provide a record of past sequences of events.

As cave streams and rivers are diverted from the passages which they formed, a whole range of different processes begin in the development of the cave. Draining out of a phreatic cave system may leave the passages completely dry, but there is usually still some water which must flow down through the limestone, and in doing so it may utilize the ready-made routeways of the older caves. The result is that the old phreatic tubes can have canyons cut in their floors by the younger vadose streams which may invade them. The relative sizes of the tubes and canyons may vary over a wide range, but the common situation of a small canyon cut in the floor of a large tube gives the classical cross section of a keyhole to the passage. Many cave systems all round the world contain passages named 'Keyhole Passage', as the shape is so distinctive. Castleguard Cave in Canada contains many kilometres (or miles) of magnificent keyhole passages; the main phreatic tube is about 5 metres (15 feet) in diameter and the canyons in the floor are narrow slots up to 20 metres (60 feet) deep.

All types of sediment—sand, clay, boulders and silt—are found in cave rivers of different sizes. In active passages they are continually on the move for they are part of the erosive mechanism of any stream or river. Where the River Web goes underground in the massive Sof Omar Caves in Ethiopia, great banks of sand and gravel

Peak Cavern in England, are phreatic in origin too. The immense system of dry tunnels in Mammoth Cave in the United States are nearly all phreatic in origin, though they are elliptical rather than circular in cross section, owing to a degree of geological influence. How then have these caves been formed? Gargantua Cave, high in the Canadian Rockies, demonstrates the answer neatly. It too consists of a network of large rounded tunnels which are almost all completely dry. It has three entrances all formed where phreatic tubes open into a great limestone cliff. Looking out from these rounded doorways, the visiting caver has a superb view, across snow-covered mountains and beneath him right down to the depths of a steep, glaciated valley. A river cascades down the valley floor, and it is quite clear that the tubes of Gargantua Cave can never have been full of water while the valley has been there, for the tubes have been emptied by the develop-

are a feature of the underground river galleries, the continued movement of the sediment making the banks and shoals distinctly different shapes after each flood season. A cave stream or river decreasing in flow has a particular tendency to deposit sediment as it loses the power to move it along, and once the water abandons a cave passage the sediment is left behind as a permanent feature of the cave. Sand and clay, especially, form extensive deposits in caves, and are commonly responsible for the pleasant flat floors in some of the world's larger show caves, which have become popular tourist attractions.

Rock breakdown and roof collapse is another process conspicuous in dry cave passages. There is a limit to the mechanical strength of limestone, and undercutting of cave ceilings naturally leads to some degree of breakdown, depending largely on the strength of the rock and extent of fracturing. Perfectly clean, unsupported cave ceilings are known in many caves to span over 20 or 30 metres (60 to 100 feet) without any breakdown at all. Thin bedded limestone on the other hand is especially prone to collapse. In the Berger Cave in France a section of the main passage is known as the 'Big Rubble Heap'. For about 400 metres (a quarter of a mile) the cave, about 50 metres (150 feet) high and wide, is floored with a mass of broken slabs of limestone. Most of the blocks are a metre or so (a few feet) across, but some are the size of houses. No one knows how deep the boulder pile is, and the roof where it can be seen is jagged and broken. The rate of collapse is, however, very slow indeed; it must have taken tens of thousands of years to form the Big Rubble Heap, and none of the many cavers who have visited the Berger has yet seen a rock fall from the roof.

It is in the abandoned caves where boulder floors and rock piles are so common, because once the blocks fall from the ceiling they are left on the cave floor free from any further erosion. But breakdown is a process integral in all large-scale erosion. Collapse in active river passages results in blocks falling into the water then to be broken, dissolved and eroded away. Large cave chambers are commonly the site of extensive collapse, and indeed collapse is frequently claimed for the origin of such chambers. A solid mass of limestone in a cave roof occupies a much larger volume, however, when it is a tumbled mass of blocks fallen onto the cave floor. So collapse does not actually form caves; it fills them. But it is important in the way that it modifies caves, and as such is a contributory factor to cave erosion, especially in large chambers.

Certainly one of the largest chambers in the world is the La Verna chamber deep in the Pierre St Martin Cave in the French Pyrenees. It is circular, about 250 metres (800 feet) in diameter, and over 140 metres (450 feet) high in the centre. Formed where the cave river descends

steeply and is now lost in the chaos of boulders on the floor, its development has almost certainly been aided by the existence of a thick layer of shale now visible high in the chamber walls. As the cave river carved the limestone away from beneath, this near-horizontal shale layer would have been a marked weakness in the rock; it would have hastened the block collapse of the unsupported bed below it, and since then the rock has broken more irregularly above it to leave a relatively stable arched roof. In a similar fashion, Britain's greatest cave chamber, in Yorkshire's Gaping Gill Hole, has developed under the influence of a great inclined fault in the limestone. The chamber is just under 150 metres (500 feet) long and around 35 metres (110 feet) high and wide, with its massive sloping roof aligned on the fault plane. Undoubtedly the chamber is situated where it is, and as large as it is, because of the fault, but it has been formed by

Above : Huge twin columns loom above a terraced crystalline floor of calcite in River Cave at Wombeyan in Australia

the powerful waterfall which still hammers down into it, and collapse has just helped it.

Roaring waterfalls, massive collapse chambers and rivers completely flooding great tunnels are part, but not all, of the pattern and process of cave development. The part of cave development which is hardly seen by anyone is that taking place in the tiny narrow fissures in the limestone, and furthermore this part is disproportionately important because it is the first, initial phase of the development of any cave. The outline pattern and framework of a cave system is basically established in the first phase: the succeeding processes, the spectacular river erosion and the collapse, normally only enlarge the cave.

Most reasonably strong rocks, such as sandstone, limestones or granites, contain networks of fractures and joints, formed by the stresses of earth movements over long periods of geological time. Once these rocks are exposed to the surface,

rainwater flows into and through their fracture systems. The movement of this groundwater is extremely slow; it may take many years for water to flow right through a mountain of sandstone or granite because the fractures are so very narrow—a matter of thousandths of a millimetre once below the surface weathered zone. Such slow-moving percolation water is not capable of erosion, except in the special case of soluble rocks such as limestone. Then, even with very slowly moving water, erosion can take place—by chemical solution. Consequently the fissures in limestone, and not in sandstone or granite, are opened up from the fractures initially formed in the rocks tectonically (by large-scale earth movements).

This initial phase of fissure opening, from widths of about a thousandth of a millimetre up to about 10 times that, is by far the most time-consuming in the development of a cave, involving thousands of years. As the fissures grow wider,

erosion slowly speeds up due to the greater movement of the water, and furthermore the rate of increase of erosion also increases with widening of the fissures, in turn because of the greater water movement. Initially the water can be envisaged as percolating in thin sheets along a wide front in any single fissure. But in time some parts of the fissure become wider than others, gaining a hydraulic advantage and taking an increasingly large share of the total flow. Where the flow is concentrated, the water, which still fills the fissure, carves out little phreatic tubes—maybe only a few centimetres (or inches) in diameter. These tubes comprise a network of anastomosing channels—meaning that they branch and loop in very complex patterns which are dictated by the vagaries of the initial flow lines. Between the little tubes, the fissures in the limestone are left relatively narrow, as their water is captured by the more efficient channelways of the wider tubes. These less-developed fissures are relegated to an almost permanent role of transmitting tiny amounts of percolation water; at this early stage they have in most cases already lost their chance of developing into large cave passages.

The networks of anastomosing tubes continue to enlarge, even more rapidly as they catch larger shares of the groundwater flow and as water speeds increase in the more open channels. Less efficient tubes are abandoned in favour of smaller numbers of larger tubes. If the tubes are in a completely flooded block of limestone, such as below adjacent valley levels, they continue to enlarge until they form the massive phreatic passages so common in the world's cave systems. But where these initial tubes are formed in a block of limestone perched high on a hillside, another change soon takes place. The water enlarges one system of tubes until it is easily capable of conducting the entire flow, at first only in drier periods, so that an air surface can develop above the free-flowing water. Gravitational flow takes over from pressure flow. With the water only flowing in contact with the passage floor, a canyon cave proceeds to be incised; a vadose canyon develops from the initial phreatic phase. In the roof of many modern vadose caves shallow elliptical openings or networks of half-tubes remain as evidence of the cave's early phreatic history.

Both phreatic and vadose cave systems, therefore, have the same early origins, with the pattern of their subsequent development dependent on the availability of free downhill drainage through the limestone. The precise location of the main cave passages which develop from the initial network of fissures depends upon other factors. The overall pattern is dictated by hydrology. Maximum erosion and cave development is where there is maximum flow, which is along the steepest straight line from the larger stream sink to the nearest, low-level valley floor on the base

of the limestone. On bare limestone uplands, rainwater sinks straight into the rock without collecting into surface streams. There tend to be therefore many, even hundreds, of small fissure openings and few large enough for man to enter; limestone pavements are notorious for the small proportion of cave entrances in their complex fissure networks. A cave entrance is best formed where a stream or river gathers in a drainage basin on impermeable rock and then flows onto the limestone to form a proportionately large opening into the ground. Ingleborough Hill in the Yorkshire Pennines has a summit mass of shale and sandstone resting on a wide plateau of horizontal limestone; streams drain off the shale all round the plateau and sink into the limestone, and hundreds of cave and pothole entrances break the surface. In contrast the limestone plateau of the Derbyshire Pennines lacks a shale cap; nearly all the drainage is underground, but

via dispersed seepage, and few caves are known. The extreme case is found in tropical areas where great rivers disappear into massive cave entrances on their way through limestone hills.

The overall pattern of cave development is established to a considerable extent in the early phases of development by the very slowly moving percolation water. Without the erosive power of turbulent rivers, this water flow is especially susceptible to control by geological features of the limestone; and once established the geological influence on the pattern is difficult to obliterate. Fractures in the limestone, joints and faults, provide ready open routes for drainage. Water flowing along such fissures tends to open up cave passages which have a cross section like a vertical lens, though in more mature caves the shape can be a vertical ellipse or even only a slightly distorted circle. Such fracture-guided passages are very prominent on almost any cave map. Dowber Gill

Right : These stalactites
in Withyhill Cave have
a sparkling appearance
—a characteristic
feature of cave
decorations that are
still in the process of
being formed
Below right : These
helictites in GB Cave
in the English Mendips
look like a cluster of
twisted plant roots—
one of the reasons why
helictites were once
thought to be organic

Cave in the English Pennines is an extreme case ; the main passage is around 20 metres (70 feet) high, mostly less than a metre (3 feet) wide and nearly arrow-straight for over 1500 metres (just under a mile) without any tributary passages. Almost the entire system of passages in France's Bramabiau Cave is controlled not by one fissure joint but by two sets of joints ; the main active passage is therefore a tight zigzag of spectacular, high-rift passages.

Joints and faults are also important in guiding the water vertically down through the limestone ; so many of the world's large underground shafts and waterfalls can reveal origins in a tectonic fracture in the rock. Greece's Epos Chasm, high in the wilds of the Pindus Mountains, is formed at the intersection of two major joints. The cave plunges nearly 500 metres (1500 feet) down a series of huge vertical shafts in a total horizontal distance of less than a thirtieth of the depth.

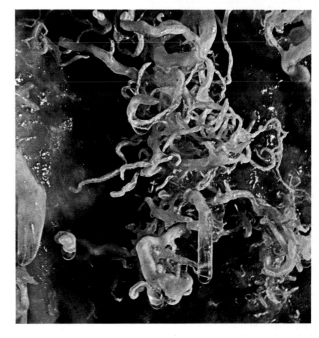

The second major geological control of cave patterns is the bedding of the rock. Limestones are sedimentary rocks, laid down in horizontal layers, or beds, on the floors of ancient seas. Irregularities in deposition, which occur quite naturally, account for slight chemical variations between different beds, and even layers of clay, shale or other impurities deposited between the beds. Such features influence the solution rates during cave development and water is guided along the bedding planes, trapped in place by impermeable layers or less soluble beds. In the Gaping Gill Caves in the English Pennines, there is one passage which is a classic of bedding-plane cave development; Hensler's Passage is 400 metres (a quarter of a mile) long, averages three metres (10 feet) in width and is rarely more than half a metre (2 feet) high. On a larger scale, practically all the main passages in America's Mammoth Cave are in one bed of the limestone. Bedding

planes in the limestone are also frequently visible as the roofs of vadose canyons; they were where the cave started, and as the canyon has been cut down the roof has been left as a flat bed, or a very shallow half-ellipse.

Flat-lying limestones, cut by vertical joints and faults, therefore lead to the development of distinctive staircase-type profiles of cave systems. The great cave of Michele Gortani has been carved out of the rock of Mount Canin, in eastern Italy, in virtually six great steps—descending to a total depth of 900 metres (3000 feet). An even greater depth is reached in France's Berger Cave, but there the limestone slopes gently down and the depths are reached by a sloping passage with far fewer large shafts than the Gortani. Steeper still is the André Touya Cave in the French Pyrenees. Its main passage reaches a depth of 500 metres (1600 feet) in less horizontal distance, and with only two small shafts, because the

limestone beds are so steeply inclined. This steeply plunging canyon runs along joints which also guide it down the bedding, and it only finishes where the water intersects a particularly large fracture zone; there it descends a single immense shaft 300 metres (nearly 1000 feet) deep, before flowing into another steeply descending fissure canyon.

That the early development of caves is so influenced by the geology is indicated by the distribution of anastomosing sets of miniature tubes seen in the walls and roofs of larger cave passages. Most of these proto-caves occur in bedding planes or joints, and are, therefore, well revealed when slabs of rock break away into an adjacent passage to expose a set of tubes not subsequently enlarged into true caves. But in some caves these intersecting tubes develop in a three-dimensional maze; the result is limestone looking like a sponge. A fine example of spongework has been

christened the Boneyard—after all its weird shapes—in a corner of the Big Room in America's Carlsbad Caverns. It indicates how much more there is to the geology of limestone than bedding and joints; indeed grain size in limestone and the magnesium content of the rock are just two more of the many factors which influence cave development in ways which are not yet fully understood by cave scientists.

Caves must therefore be seen as complex features developing under geological, hydrological and topographical influences. Not only do these factors determine where caves are developed but they also control the pattern and structure of cave systems and even the shape and nature of the individual cave passages. The geology of the limestone is primarily responsible for the pattern and framework of the cave development, while the hydrology both above and below the ground has a rather more basic

influence on the broader distribution of caves. The topography of the surface (both past and present) mainly in the depth and pattern of the valleys is the principal factor determining the distribution of vadose and phreatic passages within the cave systems. To understand the development and formation of caves, or even a single cave passage, it is therefore necessary to understand the relationships with all these factors—and while it is clear in some cases, it poses many problems in others.

Some caves are easily understood. Mammoth Cave in the United States is famed for its large, long, dry tunnels. They are elliptical in cross section—clearly phreatic and yet not perfectly developed tubes because of a degree of geological control; the ideal circular cross section has been drawn out into an ellipse because a particular bed of limestone is more soluble than those in the roof and floor. The great Italian cave of Piaggia Bella, high in the Maritime Alps, is a deep and complex system. Stream passages from three separate entrances all lead down steeply through the limestone until they unite in a single main passage which has a much more gradual gradient. This is because the water has arrived at the base of the limestone beds which are only gently inclined; the walls and floor of the main streamway are cut in a spectacular, bright green metamorphic rock, only the roof being limestone. The free-flowing vadose water which formed these parts of the cave was directed by gravity down through the limestone until the geology held it to a gentler graded course.

The causes of the patterns and changes in passage detail may be less immediately obvious in other caves. Castleguard Cave in the Canadian Rockies is noted for its spectacular examples of two contrasting passage types: magnificent phreatic tubes such as the Subway, and the narrow vadose canyons known as First Fissure

and Second Fissure. These two passage types alternate along the length of the cave's single main line. But a closer examination reveals that the phreatic tube is continuous—the remains of it are just visible in the roof of the fissures. So the apparent alternation of passage types is in fact no more than local development of the canyons—which can be explained in terms of young vadose streams invading only parts of the older phreatic tubes, not touching other parts, and remaining ponded and powerless in others.

Swinsto Cave in the English Pennines also provides an apparent anomaly in that the water which enters at the sinkhole flows northwards for over half a kilometre (1600 feet) before doing an about-turn and flowing southwards for double that distance to emerge from the spring of Keld Head. Yet a simple study of passage shapes reveals at least part of the quite complex development behind this odd pattern. The northward-orientated passages are vadose canyons, and the gravity-controlled flow of water is therefore down the slope of the limestone beds, in this case towards the north. But the southward-orientated passages were originally phreatic tubes (only some are now partly drained) which were guided by the laws of hydrology to the lowest point where the nearby valleys cut into the limestone, here to the south. So the great turn in the path of the Swinsto Cave stream may be explained; but there are features of caves which are not so readily understood.

The main passage in England's longest cave system, Easegill Caverns, is high, wide and relatively straight for nearly a couple of kilometres (over a mile)—except that half-way along its length it contains a series of narrow zigzags known as the Minarets. The Minaret passages are controlled by a cross-cutting fracture zone in the limestone but it is not clear why they are so much smaller than the rest of the main passage. The

Left: The brilliance of this calcite curtain makes it the highlight of a decorated chamber in Shatter Cave in the English Mendips

recently explored Martin Hollow Cave, in the Boston Mountain Escarpment of Arkansas, poses an almost similar problem. A spacious main passage, 6 metres (20 feet) high and a little wider, carries a stream for a kilometre (over half a mile), but suddenly ends where the water turns into a narrow zigzag canyon passage 100 metres (300 feet) long leading directly to the exit. Why the change in passage style? It could be that the smaller passage is younger, or perhaps it is purely vadose while the main passage has a period of phreatic development, or perhaps the narrow canyon is cut in a less soluble bed of limestone. The latter seems to be the most likely answer but this will only be confirmed by careful mapping.

It is a feature of every geography textbook that springs frequently occur at the bottom of beds of limestone, due to the groundwater draining down as low as it can and then flowing out to the surface on top of the underlying impermeable

rock. Chapel le Dale valley in the English Pennines is a classic example of this, with dozens of springs all along its edges at the base of the nearly horizontal limestones. One of these springs is at White Scar and is special because the cave behind it is known for nearly three kilometres (two miles). At the spring the cave has a limestone roof and a slate floor—just as if it were straight out of a textbook—but this only lasts for about 100 metres (300 feet); for the rest of the cave's length, upstream, it is developed at higher levels entirely in the limestone. The reason for this is not yet known, but it is clearly an oversimplification to visualize groundwater collecting at the contact of limestone and underlying impermeable rocks.

Of all the mysteries associated with cave development and erosion, it is the formation of big cave chambers which remain among the most thought-provoking. Why is there a big chamber in one spot and not the next? Even small cave

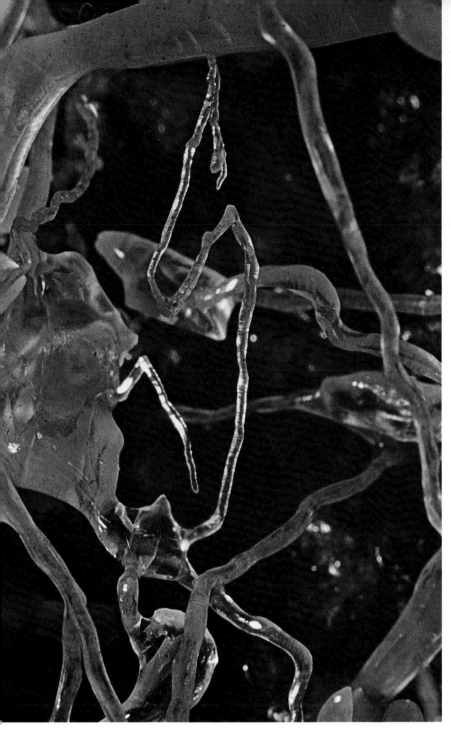

Above : The intricate pattern of the helictites in Temple of Baal Cave at Jenolan, Australia, frequently resembles a spider's web

role—they both initiate caves, and help to fill them.

As a network of fractures in limestone first begins to transmit water, some fissures open into sizeable caves and capture much of the flow from the undeveloped fractures. The latter are demoted to the permanent task of carrying no more than tiny amounts of percolation water—which have a vital role, however, in the realm of cave development: meanwhile, the main drainage channels are enlarged, until under vadose conditions air-filled caves are left above the subterranean streams. Into these caves then drips the percolation water from the truncated ends of the still small fissure openings. Much of this percolation water has seeped through soil into the rock and, charged with carbon dioxide from the rich soil air, it dissolves high contents of the limestone in its tiny fissures. Only when it reaches a large cave does it again come in contact with air. The low carbon dioxide content of normal cave air means that some of the gas diffuses out of the percolation water till both air and water are in equilibrium. The direct result of this diffusion of carbon dioxide from the water is the precipitation of the calcium carbonate, which is in solution in the water but yet entirely dependent on the gas for its solubility.

Stalactites and stalagmites are a feature of almost every cave, the former hanging from the ceiling, the latter standing up from the floor. Both are most commonly made of the mineral calcite—calcium carbonate—the same material as the limestone, which it is in reconstituted form. There are other minerals which form important cave deposits, but excluding sand and mud calcite is by far the most common material of cave decorations—the beautiful white deposits which adorn caves all round the world. Calcite, together with all the other minerals, is just deposited by dripwater, seepage water and film-water clinging to the cave walls. In mature cave systems water falls into two categories with totally opposing functions—the main streams and rivers erode the caves more and more, and the mineral-saturated drips of percolation water fill the caves with new and beautiful deposits.

Stalactites and stalagmites form almost hand in hand. Water seeps down a joint in the limestone till it breaks out and hangs as a drop on a cave ceiling. It adjusts itself slowly to chemical equilibrium with the cave, and the direct result is that it precipitates calcite on the rock where it hangs. But this is a slow process and it may fall to the floor still oversaturated with mineral, in which case it deposits more calcite on the floor. The relative numbers of stalactites and stalagmites in a single cave are, at least in part, dependent upon seepage-rates as well as saturation levels. And changes in flow-rates together with the shape of the original cave passage are jointly responsible for the great variety of shapes of the

passages may open into quite large rooms, only to continue as small as before. Is it just chance that has allowed the Big Room to develop to such a size in the Carlsbad Caverns in the United States? This enormous room, 100 metres (300 feet) high, twice as wide and 1250 metres (4000 feet) long, is the largest of the large caverns in the now barren Guadalupe Mountains. It is difficult to estimate how much geological factors, its age, and past solution rates have influenced the formation of the chamber, and one always wonders if there is another bigger one still to be found.

The Big Room of Carlsbad has developed a long way from the fissures in the limestone from which it all started. But the massive stalagmites which decorate the Big Room's floor are evidence of how microfissures are still playing their role in the development of the cave as it is seen now. For these tiny openings in the rock have a dual dripstone—as stalactites and stalagmites are sometimes collectively known.

A steady drip of water leads to the development of stalagmites with smooth profiles. If the water only has a small amount of calcite to precipitate on the stalagmite all the deposition may take place very near to the top as soon as the water lands; then once equilibrium has been reached the chemically inactive water just flows harmlessly down the rest of the stalagmite to the floor. This results in tall, slender stalagmites, such as the impressive examples in the Bamboo Grove in the Demanova Caves of Czechoslovakia. Each stalagmite in this chamber is like a slightly gnarled tree trunk, without any taper to its rounded tops, and twice the height of a man. In complete contrast a conical stalagmite is formed where the dripping water contains so much calcite that it continues to deposit, as it flows down the surface, all the way to the floor; this way the stalagmite grows as a series of cones and can reach massive proportions. Stalagmites of this type, or bosses as they are sometimes called, are often descriptively termed the 'Beehive'. More complex types of stalagmites are formed when solution and drip-rates change with time. In that way narrow towers can be deposited on top of conical bosses, or the weird tree-like stalagmites of the Aven Armand in France can develop. The single chamber in the Armand pothole contains an incredible 'Virgin Forest', formed of hundreds of tall stalagmites each resembling a column of ice-cream cornets stacked inside each other; the world's tallest stalagmite, over 30 metres (100 feet) high looms out of this fantastic forest. And changing conditions also account for the great tiered cascading stalagmites which are such a feature in both the Carlsbad Caverns in the United States and Postojna Cave in Yugoslavia.

While stalagmites can grow to enormous sizes, they can never be matched by stalactites because of the relatively low tensile strength of calcite—there is a definite limit to how much can hang down from the ceiling. A drop of water hanging from a cave roof precipitates its excess calcite where its surface is in contact with the rock, in a circular ring around its edge. The water-drop then falls off, and the next forms hanging from this rim of calcite, and so the process continues. The result is a fragile, straw stalactite—so named because it has the proportions of a drinking straw with thin calcite walls and the diameter of a drop of water. Nearly transparent or a brilliant opaque white, or even stained orange by iron salts, cave straws can occur in clusters of thousands, with each one 2 to 3 metres (5 to 10 feet) in length—though occasionally they can be even longer. Cave straws grow from water flowing down inside them, but when thin films of water also flow down the outsides of stalactites, more layers of calcite may be deposited to form the more massive varieties of tapering stalactites. There are few caves without any stalactite decorations, but there can be few caves which can be called 'decorated' when they contain only one stalactite. This is the case with Pollanionain however, a short cave in western Ireland whose single main chamber has walls of nearly black limestone and mud; these only increase the visual effect of the great sparkling white stalactite nearly 12 metres (38 feet) long hanging down from the centre of the chamber roof.

Vertically hanging stalactites are easily explained, but not so the random and weird shapes of helictites. These are small calcite formations which grow out in every direction from cave walls and from other stalactites. Like cave straws they are deposited by water which flows through them, but in helictites the internal tube is so tiny that capillary action can move the water downhill or uphill. So the helictite grows in a style guided

Below: Thousands of cave pearls lie in a series of shallow, dried-out gour pools covering the floor of a passage in Jackson's Bay Cave, Jamaica
Below left: The dried-out pools in Lost Johns Cave, England, are lined with sharp crystals of calcite

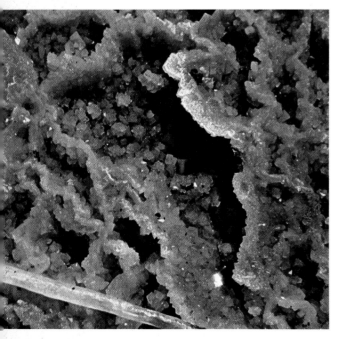

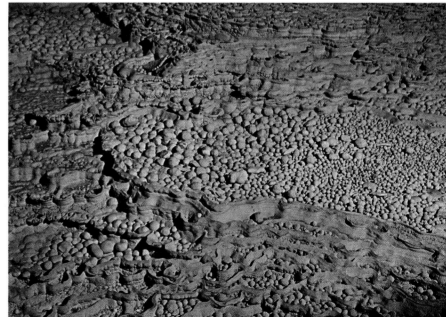

by the whims of hydraulics, by crystal defects in the calcite and even by air currents in the cave, but not by gravity. Timpanagos Cave in Utah and Moulis Cave in the French Pyrenees are just two caves containing thousands of twisted, contorted and interwoven helictites.

It is not only stalagmites, stalactites and helictites which decorate caves; a host of other shapes can be built up too by deposition from dripping or flowing seepage water. Floor and roof deposits may grow until they meet each other and cement together as a white calcite pillar or column, like those, for example, in Columns Hall in the Welsh cave of Ogof Ffynnon Ddu. Water dripping down edges of cave walls can build great hanging sheets of calcite in curtain formations; a complex-shaped wall can be covered in a calcite cascade—a shining white 'frozen waterfall'. The calcite still being deposited as seepage water flows away over the cave floor, builds up layers of flowstone.

The last stage of cave calcite precipitation is in pools of water, and two completely different types of deposit can occur at this stage. Most common are the rounded masses of calcite which can completely line pools of saturated water. When the pool is drained out by some erosive change the appearance of this type of deposit is best described by the American term cave popcorn. Whole cave chambers can be recognized as having once been filled with pool water by the lumpy calcite over their walls. The oddest effects are where stalactites have been submerged in a temporary pool, so that the popcorn grows on them and covers their smooth outlines almost like a fungus. Some of the rooms in the Carlsbad Caverns are nearly completely covered by popcorn encrustations—on walls, ceilings, stalactites and stalagmites. But even more spectacular is the calcite deposited in pools of saturated water on grains of sand which are continually moved

around—mostly by the disturbance due to drips landing in the pools. The calcite then builds up concentric layers, evenly all around the grains, until cave pearls are formed. Some are as small as pinheads but others are the size of small plums, brilliant white and often perfectly spherical.

Pools of seepage water commonly have their outlets in the form of thin films of water flowing over the barriers which hold back the pools. And within these thin sheets of overflow water evaporation and diffusion can proceed at maximum rates, so encouraging more deposition of calcite, and thus the water virtually builds its own dam. Consequently such pools, known as gour pools, can grow to spectacular sizes. A fine small-scale example, a veritable staircase of calcite-dammed pools, reaches for over 10 metres (30 feet) down a sloping wall in the Skocjan Cave in Yugoslavia; even more impressive are the great series of gour pools each with overflowing dams 5 or 10 metres (20 or 30 feet) high which extend for hundreds of metres (or yards) in France's Berger Cave.

The insides of gour pools may be lined either with rounded popcorn calcite or with calcite crystals. Calcite is renowned among minerals as one which can form a great variety of crystal shapes. From thin plates to razor-sharp pyramids, the calcite in crystal pools comprise some of the most delicate cave decorations. The Demanova Caves in Czechoslovakia contain some fabulous

Aragonite forms the most delicate of cave decorations. Where they cover rock walls, the needle-sharp crystals sparkle in the light. Above: Moulis Cave, France. Left: Chevalier Cave, Australia

examples of crystal-lined pools, some with fragile films of calcite even growing out over the surface of the still water. And Jewel Cave in the Black Hills of Dakota includes a whole series of passages with walls lined with calcite crystals, formed long ago when the whole cave was submerged in nearly static seepage water.

Even more delicate than the calcite crystals in caves are the crystals of another mineral: aragonite. Formed of calcium carbonate, chemically identical therefore to calcite but with marked differences in the atomic structure, aragonite is much rarer than calcite—but where it does form crystals it is incredibly beautiful. It does not generally grow in pools; its crystals grow from surface filmwater on the walls of air-filled caves. Moulis Cave, in the Pyrenees of France, contains a series of chambers which are fabulously decorated with aragonite. Needle-like crystals, some branching quite complexly, grow out three to five centimetres (up to a couple of inches) and encrust both the walls and a whole range of older calcite stalactites and stalagmites.

Certainly Moulis is a fabulous cave; carved out of the limestone by long-gone rivers it now contains only a small stream, but its chambers have since been decorated first by calcite drip formations and then by the galaxy of aragonite crystals. Maybe it has more than its share, but in many ways Moulis is typical of the rather remarkable world of caves.

Crystals of aragonite coat the stalactites in Moulis Cave in France (above) and Buchan Cave, Australia (left)

America, notably Venezuela and Trinidad, a number of caves are inhabited by the guacharos, often known as oilbirds. With a wingspan approaching a metre (about 3 feet), the guacharo is a normal-looking bird, but it nests deep in the dark zone of the caves. It has eyes for its outdoor feeding, and also an echo-location device similar to the bat's for underground navigation. The guacharo's sonar apparatus works on high-frequency clicks within the human audible range, but, again as with the bat, a visitor to a guacharo cave only tends to notice the frantic shrieking noise the birds make when disturbed. In complete contrast to the guacharo, the North American packrat seems to rely on memory alone to find its way around caves. It does not restrict its venture to the entrance zone of twilight, but explores considerable distances into complex systems of caves. The packrat has long whiskers and an acute sense of smell, and on its first visit to a cave must search out a safe path very carefully. But once this is done, it can retrace its path at full speed relying on memory and the trail of its own scent.

While darkness is a perpetual disadvantage of the underground, caves do offer certain advantages to its inhabitants, most noticeably protection. Bat roosts and guacharo nests deep inside caves are, at least, completely safe from normal predators, and this favourable aspect of caves accounts for the underground habits of a whole range of cave dwellers and cave visitors from the animal world. In North America the cave crickets are regular daily visitors to the caves where they rest each day between their nightly forages for food, and the harvestmen—or daddy-long-legs as they are known in Britain—are annual visitors hibernating each winter in the relative warmth of dry caves. In the Far East the salanganes, or the cave swiftlets, regularly build their nests in the dark zones of caves. In Java, Sarawak, Malaya and China there are many caves containing salanganes, and their nests are valued and collected by the local people, especially the Chinese, for they are built largely of the birds' saliva—and three or four of them can be used to prepare a bowl of the well-known delicacy, bird's nest soup.

Caves also offer refuge to even larger animals. Bears were frequent visitors to caves during the ice ages, and many must have hibernated underground as this would have been the warmest place during the winter. Cave bears are now only known from their skeletons, but the huge numbers of bones in some caves indicate that not only did many bears live and die in caves, but also that they ventured quite considerable distances into the darkness. Karlshöhle in the hills of the Schwabian Alb, Southern Germany, is a cave famous for the many well-preserved bear skeletons that have been found in its chambers.

Below : A blind, white cave cricket finds its way around the underground darkness with the aid of long, probing antennae

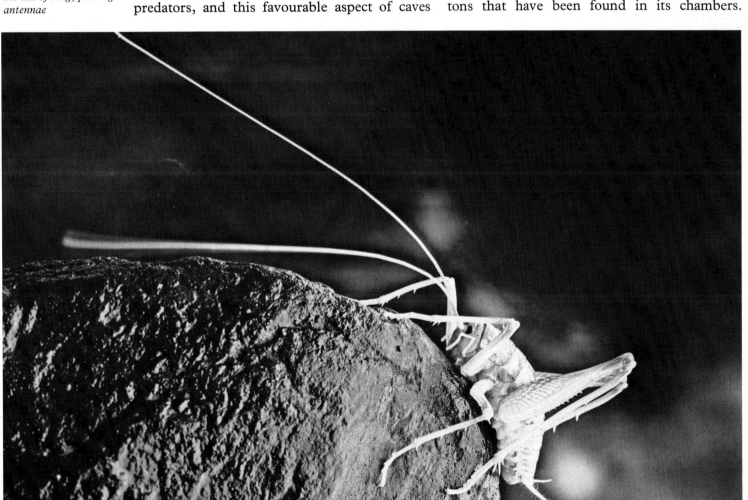

Another regular ice age cave visitor was the hyena, although it still lives today, and still uses caves, in Africa. Its skeletons, and the remains of carcasses it dragged in to feed on, can be found in cave deposits in Britain and Europe; Hyena Den in the Mendip Hills, Kent's Cavern in Devon and Kirkdale Cavern in Yorkshire are just three English sites once used as hyena lairs. The modern equivalent can be found in many parts of Africa. The little rock shelters and lava caves of Kenya commonly contain great assortments of clean bones—sure evidence that hyenas have retreated into them to feast on hard-won trophies. Makingen Cave in the flanks of Kenya's Mount Elgon is remarkable because it is used by a whole range of animals besides the hyena. Porcupines and leopards are known to visit it, probably like the hyena mainly for the purpose of enjoying a meal dragged in from outside, and there is of course the usual collection of bats

sleeping in its roof. But, in addition, elephants and cattle are regular visitors, and they venture in to lick the crystals of mirabilite (hydrated sodium sulphate) off the cave walls; mirabilite grows naturally on the rock and has the property of keeping down worms in the animals. In their search for it both cattle and elephants are even known to climb over boulder piles and into the darkest corners of the cave's main chamber.

Makingen Cave must be one of the few in the world which attracts animals into it by providing them with part of their diet. Food is generally scarce in the underground world. Any natural environment has a definite, balanced food chain. It may best be envisaged as a pyramid with vast numbers of minute organisms acting as food to decreasing numbers of larger animals, and so on upwards to a small population of the largest, most biologically advanced types. Plants normally form the base of this pyramid, but plant growth is

Top: A colony of Greater Horseshoe Bats hibernate hanging from the roof of a limestone cave
Above left: Sleeping through the winter in its damp cave home, this Whiskered Bat is covered with beads of moisture
Above: An aroused Bechsteins Bat shows a fearsome set of teeth

dependent on sunlight and consequently the food pyramids in cave environments lack this basis. The whole food supply is, therefore, rather fragile in caves; animals have to adapt and compete for what there is—and the basis of the pyramid is only provided by material brought in from outside. Some bits and pieces—leaf litter and even fragments of animal remains, for example—are washed into caves by sinking streams, but quantitatively these do not compare with the material carried in by the regular visitors such as the bats, the crickets and the guacharos which feed outside.

Bracken Cave, near San Antonio in Texas, houses about twenty million Mexican free-tailed bats each spring. Each day they rest in the cave and each night they go out searching for food, sometimes going as far away as 80 kilometres (50 miles) from the cave. It is estimated that in a single night's feeding they catch and eat between three and four billion insects. Like most types of bat, the Mexican free-tail homes in on his insect prey in flight, using his echo-location device. Then he does not catch the insect in his teeth, but scoops it up in his wings or tail, and follows this by bending double, still in full flight, to reach down to eat it. The bats have certainly developed unusual and remarkable nocturnal feeding habits, but as far as the cave ecosystem is concerned their daytime habits are just as important. A large colony of bats hanging from a cave ceiling looks like a living rug—they are crammed against each other so tightly that the rock is invisible; all that can be seen is a mass of little furry bodies, with tiny faces peering out from between their folded wings. And from the bat rug a veritable rain of material falls to the cave floor; fleas and parasites are shrugged off their hosts, and the odd dead bat falls away, but most important are the droppings. Gorged as they are from a night's feeding, these large bat colonies produce enormous amounts of droppings and these steadily accumulate on the cave floor as layers of guano—the food basis for nearly all the permanent cave dwellers.

Guano from the cave's daily visitors—whether they be bats, crickets, salanganes or guacharos— is a rich enough material to form the bottom layer of the cave environment's food pyramid. The guano does vary slightly in composition. Most bat guano is a grey to black, fine material consisting mainly of phosphates and includes many indigestible insect remains. The material dumped in caves by the guacharo birds consists of a mixture of droppings and various seeds, for the guacharo feeds mainly on fruit and plant material, rejecting hard seeds from his cave nest. The seeds fall to the guano floor and may germinate into ghostly white seedlings which may grow a metre (3 feet) high before starving from lack of sunlight; these underground forests are a well-known feature of the South American caves populated by the guacharo. Different again is the

black, tar-like, guano left in caves by a special type of bat—the infamous vampire.

Vampire bats appear to be distinctly anti-social. If they occupy a cave, there will be no other type of bats living with them. But despite this they are not quite the terrors they have been branded by mythology and folklore. With a wingspan of just less than 40 centimetres (15 inches) a vampire can settle so lightly that it seldom wakes a sleeping animal; then it drinks its victim's blood, the sole diet of the vampires. Sharp front teeth enable it to make a tiny nick in the skin for its feast, and the loss of blood does the victim no harm. Vampires earn their reputation, however, by being major transmitters of rabies. The vampire bats live in Mexico and adjacent countries farther south, and every year thousands of cattle die from contracting rabies. The bats spend their days sleeping in caves, rarely more than a hundred to a colony, and at night feed on any animals

Above: Green plants depend on light for their existence. Consequently, they are found deep in caves only where lights have been installed along tourist paths

Right : Disturbed by the photographer, hundreds of bats fly in confusion around a chamber in the Riverhead Cave, Jamaica

Above left: Proteus, the blind, white salamander of Yugoslavia, swims in a shallow pool in the Planina River Cave
Left: The attractively coloured American salamander crawls across the mud in Sig Shacklett's Cave in Kentucky

available. Newly farmed areas in the tropical karst of Central America suffer worst from them, for cattle are commonly grazed on the sedimented plains, while the small intervening limestone hills are riddled with small caves, providing the vampires with excellent sleeping accommodation adjacent to their feeding grounds.

The guano from all these daily visitors to the caves directly supports a whole range of low forms of life which are the first of the permanent cave life. Fungi and a few other simple plants may grow on it, but of course they all lack any green colouration. Green only comes with light, and consequently the only green plants beyond the entrance zone of any cave are the mosses which so commonly grow around the lights in show caves. With the fungi and plants thrive a host of other tiny animals: millipedes and centipedes crawl over the guano banks, and flatworms and isopods swim in the pools of water themselves nearly saturated with guano. Hordes of insects and many parasites complete this next layer in the food pyramid; they are all true troglobites (permanent cave dwellers), but are dependent on the guano from the cave visitors such as bats with outside feeding grounds.

A further step up the cave's food pyramid are some of the larger and more spectacular troglobites. And these are nearly all scavengers. They will eat anything, for they cannot afford to be selective in the battle for food in the relatively sterile cave environment. They eat the guano itself, any washed-in plant debris, carcasses of dead bats, any smaller animals such as the millipedes and flatworms, and even their own kind when they die; cannibalism is second nature to a scavenger. Cockroaches and beetles are the major cave scavengers in terms of numbers, but others of similar habits include scorpions, packrats, and in many of the caves of Jamaica white crabs up to the size of a hand are found. The bats and cave birds only live in large numbers in the tropical regions, notably in the karstlands of the East and West Indies, but also in South and Central America and South-east Asia. So it is in these areas where the caves most commonly contain large guano deposits, and it is the cockroaches and beetles in particular which add the distinctive touch to the cave ecosystem. A tropical bat cave is a nauseous place where the bats whistle and squeak through the air, a continuous rain of fleas and excreta falls from the ceiling, the floor is a heaving seething mass of cockroaches and insects, and to cap it all a hideous stench rises from the guano whenever it is disturbed.

Feeding habits adjusted to the cave environment are a feature of the scavenging cockroaches, but are even more remarkable in some of the water-living troglobites. Cave salamanders are known from various sites in North America and Europe, but undoubtedly the most famous is *Proteus*, the 'man-fish' from the river caves of

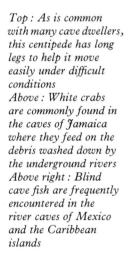

Top: As is common with many cave dwellers, this centipede has long legs to help it move easily under difficult conditions
Above: White crabs are commonly found in the caves of Jamaica where they feed on the debris washed down by the underground rivers
Above right: Blind cave fish are frequently encountered in the river caves of Mexico and the Caribbean islands

Northern Yugoslavia. Biologically *Proteus* is one of the most advanced cave dwellers, and it sits on top of the food pyramid in cave waters. Rarely more than 30 centimetres (12 inches) long, it is often called the man-fish because of its four short limbs, though it is of course totally eyeless. *Proteus*, like so many other cave dwellers, is completely white except for a touch of blood-red around the gills, but if any are taken from caves into the daylight zone they rapidly develop a brown pigmentation. Living in caves undoubtedly breeds peculiar habits, and one of the *Proteus*'s is to give birth to live young if the temperature is above 15°C but to lay eggs if it is any colder. The main habitat of *Proteus* is the cave rivers of the Postojna area in Yugoslavia, and these lack the relatively rich food sources of the tropical caves. Consequently *Proteus* is well adapted to a very sparse diet. Essentially it is a scavenger, living on almost anything—even the traces of organic debris in cave mud—but if pushed it is capable of surviving for a year or more without any food at all.

Safe in its underground world of water and darkness *Proteus* must have few enemies. It feeds on other animals but only contributes to further diets when it dies. The same cannot be said for the scavengers of the mudbanks in the caves; their life is more hazardous for they do not sit on the top of their own food pyramid. Instead, they are the food for the predators. King of the predators is the cave spider, who will prey on almost any other animal, in some cases up to the size of young bats. Spiders are among the most widely distributed members of the entire animal kingdom; there are few environments which they have not colonized. Yet they have had to adapt special characteristics to survive in caves. The overall scarcity of food means that they cannot just build a web and wait for prey to become entangled. Instead they must chase their prey, and consequently cave spiders tend to be quite large, with particularly long legs. In temperate regions, such as Europe, cave spiders are not very common and rarely larger than house spiders, but they are far more spectacular in the tropical caves. Furry individuals the size of a man's hand are common there, though even they are small compared to some of the sleek-bodied spiders of the caves of the South American jungles; they have a comparable body size, but have legs which may span half a metre (18 inches).

Caves are, and always have been, branded by many people as places to be scared of; this seems unfair, although huge spiders running out of the darkness over seething banks of cockroaches and bat-droppings tend to make the evil reputation of the tropical guano caves fully justified. While so many of the world's caves are sterile and lifeless, the ecosystems in the caves of the tropics can only be described as having developed to extraordinary, and rather horrifying, levels.

Cave Dwellers and their Art

During the summer of 1856, workmen in a limestone quarry, not far from the city of Düsseldorf in Germany, blasted open a small cave. As they were quarrying limestone, the sediment in the cave was an inconvenience, and the workmen broke into it with their pickaxes. As the debris was thrown aside some bones contained in it were broken up and also discarded, except for the skullcap, ribs and a few limb-bones which were saved. The significance of this discovery was not recognized till over a generation later, but the bones are now known to be in the order of 100,000 years old. The cave is known as the Feldhöfer Cave—a name of significance to few people. Feldhöfer Cave, however, lies in the flank of a small gorge in the valley of the River Neander—or, in the German language, Neander-thal—a name universally recognized as one of the early types of man.

It is indeed significant that a cave site should give its name to one of the most important of man's ancestors, for anthropologists and archaeologists around the world have always looked to caves for their very best material. Neanderthal man was the first to have a modern-sized brain, though his skull still looked archaic with a sloping rather ape-like brow. He commonly lived in caves, as did many of his ancestors, and his descendants too—Cro-Magnon man. Best known for the fabulous cave paintings he left behind in the limestone regions of southern France and northern Spain, Cro-Magnon man was the first truly modern man—at least in terms of his physical structure. And Cro-Magnon man, too, is named after a cave site. Five skeletons of his type were uncovered, during preparation for the foundations of a railway line in 1868, in a small cave at the foot of the hill of Cro-Magnon in the Dordogne region of France; it was only later that the cave art was discovered and recognized as his. As man evolved and developed, through the stone ages of the last million years, he lost his ape-like characteristics, became industrious and organized into interdependent communities, and at the same time developed a need for shelter. In cave regions this was most readily offered by the caves.

Caves acted as shelter and allowed man to conserve his energies for the rigours of the harsh Stone Age life. Once inside a cave he was safe from weather and predators, and it was therefore the ideal place to spend the long hours of darkness. Similarly, should he have died in his cave, the body of ancient man was protected and preserved. Buried by the accumulation of all manner of debris in the succeeding years, his remains could lie undisturbed until brought to light by chance, or by a cave-excavating archaeologist.

In the search for evidence and relics of man's activities in bygone ages, cave sediments are indeed bonanzas. Major cave systems carrying streams into the depths have never been inhabited by man—living in one would have been comparable to camping out in a river-bed prone to flood. It is the dry, abandoned caves—often quite small—which have offered shelter to man, and to all kinds of animals as well. In some cases these are associated with, or are the entrance parts of, major cave systems, but in others they are such shallow caves that they are better described as rock shelters. Sediment which accumulates in this type of dry cave is unlike any formed in the open air. The same protection which the cave affords man also cuts down the rate of sedimentation. Consequently the debris left by man—abandoned or broken stone tools, discarded bones from feasts, the bundles of grass carried in to make bedding, and even the bones of the dead—forms a significant proportion of cave deposits. A single cave in the limestone of the Lebanon had been used as a site for Stone Age tool-making for probably 50,000 years, and when excavated it

Left : Conical towers carved by rainwash in soft volcanic ashes at Goreme, Turkey, form a weird landmark. The rock is easily worked by hand and each tower has been hollowed out to form a cave house

Left: The Dead Sea Scrolls were unearthed in these isolated caves in the wilderness of Judaea, west of Jordan's Dead Sea

yielded a million pieces of flint—the tough sharp-breaking mineral used so much for axe-heads and arrow-points by our ancestors.

Man's early history was during the period of geological time known as the Pleistocene—essentially the last two million years. This was a time noted for its great climatic variations, culminating of course in the four major ice ages. Such changes in climate then rendered some caves uninhabitable. Man moved out and while he was gone sediment accumulated—fine silt blown in during cold or hot dry periods, or laminated gravels washed in perhaps during the cold wet periods as the glaciers melted and retreated. These sediments are relatively sterile to the archaeologist, but they do help to build up a layered structure in the cave deposits which enable excavated material to be related not only to man's own cultural development but also to the changing environments of the Pleistocene. The other great value of these sterile layers is that they bury and preserve for ever the often delicate remains of man and his activities.

In the collapsed caves of Chou-k'ou-tien in eastern China a sequence of deposits 50 metres (160 feet) thick has been excavated. Most spectacular of the finds were the skulls and other bones of 'Pekin man' as he was called, who lived there half a million years ago. Found with him were remains of fire hearths, stone tools and worked bones which tell so much of his way of life. Thor's Cave in the English Pennines contains a whole series of deposits, the oldest of which include reindeer bones left by the Palaeolithic, Old Stone Age, inhabitants. Above that is a succession of layers containing beakers and beads from the Bronze Age, then fragments of pottery from the Iron Age, and finally, closest to the surface, coins and pottery from Romano-British times.

Some of North America's earliest inhabitants also left their mark in cave deposits. Sandia Cave in New Mexico has yielded lance-points more than 10,000 years old, and similarly dated material has been recovered from Danger Cave in Utah. This was originally a wave-cut cave formed on the shores of the great Lake Bonneville—the inland sea now shrunken to the Great Salt Lake—but it was used by man in between inundations due to changing levels of the lake, each of which left its own layer of sediments. Though these thousands of caves which were inhabited by man all round the world yield so much to the archaeologist, there are still the exceptional cases—the few special caves which were not homesites but contain their own rare types of buried relics. For right through time, caves have offered a marvellous hiding-place for treasure. St Bertram's Cave in the English Pennines was known by the local inhabitants in Saxon times, and when the marauding Danes sailed up the Trent to sack the riches of the low-

Right : Artificial caves carved in the hillside as well as more conventional houses provide shelter for villagers in Goreme, Turkey
Below : These caves at Ani in eastern Turkey provided shelter for ancient man some 4,000 years ago

lands, valued treasures were carried to the hills for sake keeping. Too safe in this case, for even when it was recently excavated St Bertram's Cave yielded a whole hoard of coins, gold wire and silver brooches. In the same way it appears that people living in the Dead Sea area in about the third century, when the region was known as Judea, considered some of their most valuable treasures to be documents and parchments which were already historically old to them. So they were tucked away for safe keeping in the remote wilderness caves of Khirbet Qumran—and there they were preserved almost perfectly until they were rediscovered in 1947 and named the Dead Sea Scrolls.

Limestone walls rise precipitously from the plains of Kermanshah to form the rugged mountain of Kuh-e-Parau, in western Iran. A handful of houses forms the village of Bisetun, lying just on the plain where it is overlooked by the south-eastern end of the high cliffs. Only a stone's throw from the village, up a short scree slope, is a cave entrance—which, long ago, was occupied by Neanderthal man. The cave is less than 10 metres (30 feet) wide and long, but it gives shelter from the rain, and faces south to catch the warmth of the sun. It is dry, has no draughty corners and has a sand floor. A minute's walk away at the foot of the cliff is a spring which provides a permanent flow of clear water. The plain, overlooked by the cave, is now an irrigated semi-desert, but in bygone wetter climates it must have supported rich grazing land and herds of animals. To Stone Age man a good home had to offer shelter, comfort and protection, and have food and water nearby. Bisetun Cave did just that, and indeed so many caves did all round the world—so where available they were used as dwellings in preference to such frail constructions as skins stretched over saplings. The latter were used to make summer shelters for ranging hunters, but for thousands of years caves probably provided the only permanent homesites in many parts of the world.

In limestone regions such caves abound, but there could be more reasons for Stone Age man choosing or avoiding certain caves. In the English Midlands, the little gorge of a tributary to the River Poulter is cut between lines of low cliffs better known as Creswell Crags. A whole series of caves and fissures cut both cliffs, forming ideal dwellings. In this situation, the caves on the north side with an open southerly aspect catch the days' warmth; yet excavations have shown that the more popular homesites were those on the south side. The apparent anomaly is explained by the climate during the last Ice Age when the caves were inhabited; the nights were cold, and hard frosts shattered the rock, then the greater daytime warming-up on the south facing cliffs led to more rock breakdown and the cave entrance arches slowly disintegrated;

in the more stable temperature zone on the south side there was less breakdown—the caves may have been a little cooler, but they were much safer.

Caves were used as ready-made, convenient and safe shelter by man right through the primitive ages—and indeed, a few still are. The Tasaday people, perhaps the world's most primitive at present (recently discovered in the Philippine Islands jungle), live in caves in keeping with their simple way of life. In the mountains of Kenya, families are known to be living in lava caves, and some even house their cattle in the safety of inner chambers of their 'homes'. There can be few instances in the western world where people now live permanently in natural caves, but there are plenty of caves in such regions as the south of France which are still used as store-houses, both by farmers and by the manufacturers of wine. Caves which are still inhabited in the more developed countries are mostly artificial; best known are the whole villages of caves cut into the loess deposits of the Hwang Ho basin in northern China. (Loess is a kind of silt just soft enough for non-mechanized excavation, yet tough enough not to collapse across a cavern roof.) In similar fashion are the cave houses cut into the steep sided hills of volcanic ash in the Goreme region of central Turkey. Some of these houses being tailor-made can really be quite comfortable, and modern 'cave living' does not compare with the hardship of life in the natural caves during the ice ages.

Human bones some ten thousand years old, found in the Cheddar Caves of the English Mendips, were covered by cuts and deep scratches. Skulls of Pekin man from the Chou-k'ou-tien caves were broken open. It seems, therefore, that both these groups of cave dwellers were cannibals —human meat was just one more type of food. In complete contrast, some excavated caves in the Middle East have yielded masses of animal bones; the Stone Age inhabitants probably had a better diet when they lived there during the wetter periods of the Pleistocene than do the modern peasants in their semi-arid environment. In general, though, life in the caves was sparse. Such artifacts as flint arrow-heads, stone chippings and bone implements together with discarded bones, excavated from old cave deposits, offer a fair picture of cave life in the early stone ages. Also commonly found are traces of the fire-hearths which stood in the cave doorways, providing not only warmth but also a measure of protection against predatory animals. It was only more recently that levels of industrial achievement improved and the caves became more than primitive unmodified shelters.

In the American South-west the Colorado Plateau region is renowned for its many cave dwellings. The numerous canyons which cut through the Plateau are flanked by steep walls of sandstone; many of these cliffs were undercut

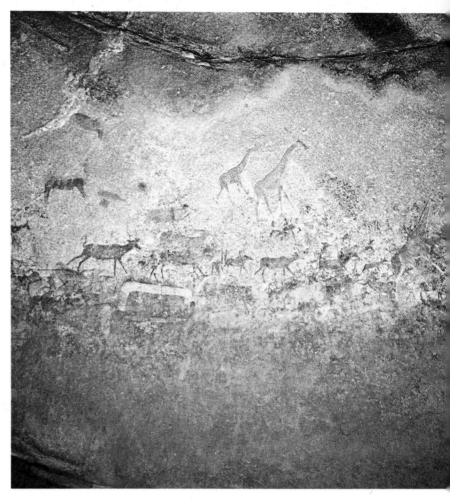

by the rivers which formed them, and the resultant niches, overhangs and rock shelters provided sheltered sites for the early Indian inhabitants of the region. Both Pueblo and Navajo Indians were responsible for these cave homes, built mainly over the last few thousand years. The caves are neither long nor deep, they are just rock shelters; but they are of sufficient width to contain not merely single homes but whole villages. Betatakin Cave in northern Arizona contains over 20 houses inside its high sweeping arch, and the rather lower, wide cave openings of nearby Mesa Verde and Canyon de Chelly are also packed with the dwellings of whole communities. Montezuma Cave, also in Arizona, is distinctive because it is not at the foot of the great sandstone cliffs, but part of the way up. Built by the Pueblo Indians about 800 years ago, the only access up the cliff is by 25 metres (80 feet) of wooden ladders, to a castle-like building of 20 rooms arranged on five levels.

Europe can barely match the grandeur of the Indian cave villages of the United States except at Predjama. Situated in the limestone heart of northern Yugoslavia, not far from Postojna, the cave at Predjama has a huge arched entrance overlooking a fertile valley. It offers a magnificent defensive site and as long ago as the thirteenth century a castle was built in its portals. After a history of attack and siege this fell into disuse, and a newer larger castle was completed in 1570.

Above: Relatively recent bushman paintings in Nswatugi Cave, Rhodesia, depict animals that roam the nearby plains
Above right: The bison painted on the roof of the famous Altamira Cave in northern Spain show the impressive effect that has been obtained by using just two colours—red and black
Right: This skeleton of a woman lay undisturbed for thousands of years in a cave at Les Eyzies, France

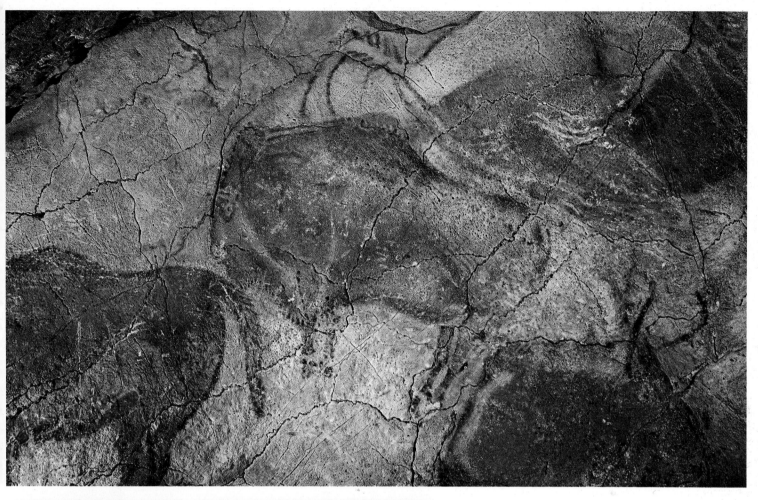

To this day, the Predjama Castle stands overlooking the valley, with its back walls safely tucked into the great cave vault.

Caves are of course superb as defensive sites, and this use is really only an extension of the shelter offered to earlier cave dwellers. But all this only involved the entrance, and there had to be another reason to attract ancient man into the depths of long cave systems. The prehistoric Indians in Kentucky, for example, lived in the entrance chambers of the many limestone caves, but also foraged deep inside Mammoth Cave and Salts Cave in search of the mineral gypsum (hydrated calcium sulphate) which grows on the passage walls. Rather different was the motive for such visits to the inner reaches of the great Niaux Cave in the French Pyrenees. Its vast entrance gives way to over half a kilometre (500 yards) of spacious passage, with some shallow pools right across its floor, and leads to a scramble over great piles of collapsed boulders and then into the 'Salon Noir'—the Black Room. A series of alcoves pock the wall of this great chamber, and they are now embellished with exquisitely composed paintings of all varieties of animals; there are over 50 of them and they are 10,000 years or more old. Niaux is just one of the many fabulous painted caves spread through the limestone of southern France and northern Spain.

The little cave of Altamira, close to the northern coast of Spain, rose to fame in 1879

when it was found to contain fine multi-coloured paintings of animals all over its roof. The paintings were dismissed as modern or fake, however, by the authorities of the day, even though the landowner-discoverer remained sure they were prehistoric; he died a disappointed man with his discoveries still unaccepted, because it was the turn of the century before numerous other painted caves were discovered and both they and Altamira were recognized as the works of Stone Age man. The limestones of northern Spain, in a belt between the main peaks of the Cantabrians and the coast, and extending from Oviedo eastward through Santander, have revealed a series of painted caves. Altamira was first and one of the finest, and the later discoveries came to a climax in 1968 at Ribadasella: the Cueva Tito Bustillo has a huge main gallery extending for over a kilometre (nearly a mile) to a chamber adorned with animal paintings of a

quality and splendour rivalling those of Altamira, and of course Lascaux.

Niaux is the best of the Pyrenean painted caves, and the third great concentration of cave art is around Les Eyzies in the limestone gorges of the French Dordogne region. One of the most distinctive caves near Les Eyzies is that of Les Combarelles, for it is the only cave site of Stone Age art without any paintings; this one is exclusively adorned with engravings of various animals; there are over 300 figures finely engraved on the walls of the 200 metre-long passage. Even paintings and engravings were not the only forms of cave adornment—Cro-Magnon art also developed in three-dimensional forms. Two Pyrenean caves are well known for their remarkable clay statues. In the Tuc d'Audoubert there is a magnificently moulded and carved pair of bison, but they are surpassed in splendour by the clay sculptures of the Montespan cave. A single

Above left : At Addaura Cave, Sicily, the Stone Age art consists of figures scratched into the walls
Above right : Most famous of all the painted caves, Lascaux Cave in south-west France has a main gallery with continuous friezes of beautifully coloured paintings on both its walls
Right : The Lascaux paintings depict many different animals, but bulls are among the commonest

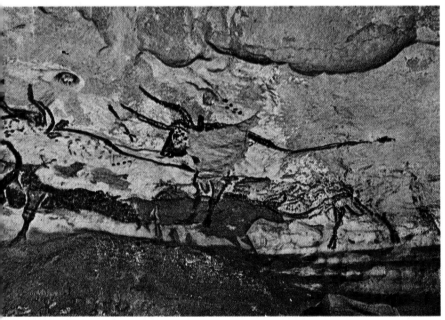

chamber in the heart of this cave system has a variety of animals—horses, bison and bears—depicted in clay; there are 30 finely executed statues some over a metre (three feet) long, and in addition numerous bas-relief mouldings and mud pictures painted on the cavern walls. Some of the statues have slumped and collapsed as their weight has told over the passage of time, but others stand beneath drips of water and have been covered with a coating of calcite flowstone, so preserving them more effectively than the original artist could ever have hoped or dreamed.

Back in the Dordogne lies the most famous of all the Cro-Magnon art galleries: the cave of Lascaux. This remarkable cave was only found in 1940 when a dog fell down the narrow entrance cleft in a wooded hillside! But that tiny opening leads straight into the main chamber, some 30 metres (100 feet) long and bearing a frieze of paintings on its upper walls. Horses, cows, bulls,

deer, ibex and even bears are all represented, and the larger of the individual paintings are some 5 metres (16 feet) long. Separate generations of painting are represented, so that individual pictures are partly superimposed, adding to the confusion in some cases but also heightening the effect of action and life. From the end of the main chamber two passages lead off, and here the paintings are even more crowded and spectacular. The passages are both keyhole-shaped, tall enough not to confine a walking man and yet with ceilings easily in reach of the artists; most of the paintings are on the smoothly rounded upper part of the passage, not on the more roughly eroded lower walls. The right-hand passage, the Lateral Passage, is the longer, and its walls also bear a host of engraving and etchings, some painted in, while others are bare or even partly obliterated by other paintings. Some way along, a shaft breaks its floor, and its depths are only accessible by a rope. Down there are more paintings, this time of bison, a rhinoceros, a man and a bird. The Lateral Passage narrows further along and then opens up into the 'Cabinet des Félins'—the Feline Chamber—again decorated, and here there are engravings of cave lions too.

How did our prehistoric ancestors produce all these works of art? The engravings were simply scratched out with sharp-edged flints, and in some caves fingers were enough to carve an impression in soft or mud-coated walls; Montespan Cave in France and Koonalda Cave in Australia both bear primitive finger-paintings. The real works of art, such as at Lascaux and Altamira, were done by the Cro-Magnon artists combining a fair range of colours with a considerable degree of artistic talent.

Various compounds of iron occur naturally in cave sediments, and provide a range of reds, browns and yellows (the ochres) for the artist. Naturally occurring manganese oxides provide a dense black, though this colour could also be made from soot. If vegetable dyes were used they have not stood the test of time, so this limits the cave art colour range. The coloured mineral powders were then mixed with water in any convenient hollow receptacle—the crowns of human skulls were used on more than a few occasions. Application to the walls may have been just by the finger, but the more refined artists used carefully frayed ends of sticks to paint the fine lines, and then pads of lichen or moss to colour in the larger areas. Another technique was to blow the paint on to the walls through hollow bones, and this is most clearly seen in the many examples of silhouette hands.

The 'painted hands' of Gargas Cave in the French Pyrenees are best known because they contain a number of mutilated examples—fingers too short or missing—and one wonders if leprosy was rife among these ancient artists' colonies; similar but unmutilated hands are also found in caves in Spain, besides caves farther afield in America, Africa and Australia. The real skill of the cave artist is demonstrated where he has used the natural form of the cave to pick out his animal paintings: the bulls in Altamira Cave, each painted over a natural bulge on the ceiling, display so well ancient man's feeling for his art.

Although France and Spain contain the best examples of cave paintings with their many Cro-Magnon art galleries, they do not have the monopoly. Each part of the world has its own cave art, though most is much younger than that of western Europe. Russia has its Kapova Cave in the Urals with paintings of the mammoths that roamed the area during the ice ages. Both American continents have their own cave art such as the Indian paintings which abound in the sandstone rock shelters of the western deserts of the United States and the ancient Inca

Above left: The little rock shelters at Tassili in the deserts of southern Algeria contain paintings of animals that can no longer live in this arid region
Above: As well as animals, men and women with Negroid features can also be discerned in the remarkable Tassili Cave paintings
Right: Aboriginal paintings decorate a small rock shelter in northern Queensland, Australia

Left: The famous caves at Ajanta in India consist of shallow natural openings in a river cliff which have been enlarged artificially to house a series of spectacular temples
Below: Inside the temples of Ajanta brightly coloured paintings decorate the rock walls

paintings in the Huagapo Cave high in the Peruvian Andes. Australian cave art ranges across the continent: finger paintings are known from Koonalda Cave in the Nullarbor desert, and the little rock shelters around Ayer's Rock in the very heart of the country contain many simple paintings; the more advanced art however, is in Arnhem Land in the north where the caves offered permanent homesites for the early inhabitants in climates more amenable than those of the deserts.

Cave art in Africa covers a considerable length of time. Probably the oldest is found in the rock shelters in the walls of gorges cutting through the plateau of Tassili-n-Ajer in southern Algeria. The paintings there exhibit an excellent feeling for movement, and they are accompanied by various hieroglyphics which are not yet understood. They are thought to be 5000 years old, and depict life as it was long before the area was

57

the desert that it is today. These cave paintings appear to have been made in a sequence which reflects the climatic deterioration of the region; buffalo are shown in the oldest paintings, followed by giraffes, then cattle accompanied by very negroid-looking humans, and finally horses and even camels. Farther south, especially in Rhodesia and South Africa, the cave paintings are complex in design and much younger—extending to modern times. A cave in the Serengeti Park in Kenya has revealed Masai paintings which include cattle, lions, elephants, people and even motor cars!

Why then does primitive man paint in caves? In the case of the Serengeti and Tassili caves the paintings appear to be a means of recording life at the time. But there must have been more reason than that for Stone Age artists to venture so far into the Niaux Cave, or even down the shaft in Lascaux. Cave art must have had a ceremonial or religious significance, possibly connected with the ability to conjure up spirits as required. Many paintings show hunting scenes, some with wounded animals. The clay statues of Montespan are pitted with spear-marks and one statue has been beheaded; footprints in front of them confirm that they were used in a form of ceremony, perhaps before a hunting party set forth for food. Some pictures of humans have grossly exaggerated sexual organs, and the same applies to the rotund form of the Venus carved out of limestone and found in the Willendorf Cave in Austria; fertility cults also had their place in cave art. This hidden meaning in cave paintings has been demonstrated by an aborigine seen at work in an Australian cave. He did not just compose his paintings; in between each stroke of his brush he stamped and chanted his way around his own particular cave; certainly to him cave art was more than just art.

Being beneath the ground, caves have a head-start as sites of mystical significance, when the earth is a feature tied to the very roots of so many of the world's religions. Sacred and holy caves exist all round the world, some now only relics of the past, but some still the objectives of great pilgrimages. And it is not surprising that so many cave sites are associated with the mystical religions of the East.

It is firmly believed by the local people that a great saint named Gupteswary, some time in the past, visited a cave near the little town of Kusma in central Nepal. The cave has now been named after the saint, and the foundation for this story is the idol of the deity Shiva which Gupteswary is supposed to have created in a chamber deep inside the cave. The idol is in fact a very impressive stalagmite some 3 metres (10 feet) high, by far the largest in this quite small cave which is now regarded as so sacred that it recently warranted a

special visit by the king of Nepal. An important symbol and indeed a recurring theme in Hindu religion is the lingam—the phallus; and the phallic proportions of so many stalagmites therefore make them popular objects of worship in caves throughout the Indian sub-continent. A limestone cave near the town of Dowlaishweram not far north of Madras is doubly sacred to the local Hindus. Firstly its inner reaches are supposed to extend to the holy city of Benares—which is a geological impossibility—and secondly it contains a stalagmite lingam which is visited by women who desire to have children.

Sacred caves are not restricted in India to Hindus, for the second great religion of the country, Buddhism, also has its own holy caves. Best known are those of Ajanta in the western state of Maharashtra. Twenty-nine caves open into the foot of a great curved river cliff cut in sandstone and, though shallow rock shelters may

59

Left: Precipitous limestone walls ring this natural sacrificial well at Chichén Itzá, Mexico

have existed originally, all the caves are essentially artificial. They were cut between 200 BC and AD 700 and include among the living quarters four chapels with their rock walls and pillars carved and painted to depict the life of Gautama Buddha.

The triple combination of caves reaching deep into the earth, the holy stalagmite lingams, and the sacred mountains of the high Himalayas cannot fail to attract the devout and religious. Halesi Cave in the foothills south of Mount Everest is the regular scene of religious ceremonies in its great main chamber, but most revered of all the Himalayan sacred caves is Amarnath. Situated high in the snow-capped limestone mountains of Kashmir, Amarnath Cave has only a single chamber open to daylight, but in it is a lingam—a massive ice stalagmite. This stalagmite is the object of annual pilgrimages held each June or July, when thousands of devotees make the

arduous two-day march from the nearest village. Many go in hope of cure from illness, and the long walk and harsh mountain weather take a terrible toll among the weak and sick irrespective of alleged healing powers.

A rather different kind of religious ceremony used to be performed at the cave at Chichén Itzá, the Mayan city in the Yucatan peninsula of what is now Mexico. The Maya people had their great city at Chichén Itzá around the time AD 500 because the site included two great cenotes—deep potholes about 30 metres (100 feet) wide by 30 metres deep; they are open to the sky, and at the bottom lie deep lakes. In the waterless limestone plain of the Yucatan these natural wells offered the only water supply, and at Chichén Itzá one was used as a sacrificial pit. Young men and women were thrown to their deaths, accompanied by offerings of gold and jade. Some of the precious relics have since been found on the

lake-bed, and there is perhaps more treasure still lying in the depths.

In almost similar fashion, caves in Europe have been associated with the offerings and dead of ancient peoples. Victoria Cave in the English Pennines was a votive site in Romano-British times, when the inhabitants of the area would throw coins in through a hole in the roof of the cave chamber. Farther south in the Mendip Hills the roomy entrance chamber of Aveline's Hole saw a long period of use as a residence before it was sealed up 6000 years ago—with a hundred bodies laid out on the earth floor! The cave of Teshik-Tak in Uzbekistan, Russia, had some special significance to Neanderthal man, for it contains the shallow grave of a child surrounded by six pairs of horns of the Siberian goat all pointing inwards and down. In Switzerland, the cave of Drachenloch contains evidence of a Stone Age bear cult; seven neatly laid out bear skulls are all that remain of a bygone ceremonial burial.

Pride of place among Europe's sacred caves must surely go to the youngest, and one of the few regarded as holy by modern, western man— Lourdes. Lying in the foothills of the French Pyrenees, the sacred shrine of Lourdes is only a small rock shelter opening at the foot of a low cliff. But it is believed that there the Virgin Mary revealed herself within the shadow of the cave to a young peasant girl in the year 1858. Since then Lourdes has been holy to all Roman Catholics, who believe the grotto has miraculous healing powers. The sick and crippled go there on pilgrimages of faith, and a hospital is at hand only a short distance from the cave to maintain those who need to wait for a cure. This tiny grotto now receives about two million visitors each year—an impressive testimonial to man's long-standing relationship with caves.

The Study of Caves

Long ago man used to visit caves for shelter and protection, for a cave could provide a ready-made roof over the head of Stone Age man and his family. Then he progressed as his skills developed; he could build a home more comfortable than that offered by the chill of the underground. So for a long period of his development, man lost interest in caves. This phase lasted in Europe through most of the Middle Ages. Events then went full circle and man has now renewed his interest in caves, but for very different reasons. The economics of modern civilization have changed man in numerous ways, not least of which has been his increased amount of time free from the necessary routines of survival. This time-factor has involved man in more leisure activities, and in more research into the unusual facets of the world: caves have offered him both.

The subterranean world of caves is one of just a few environments where sport and leisure are inextricably mixed with science and research—out of necessity. Cavers, potholers, spelunkers—as they are known in the United States—speleologists: by any name they are all people who enjoy exploring caves, except that the word 'speleologists' implies a little more. Speleologists are the cave scientists, who venture underground to study either the patterns of water movement in the limestone, or the animals that live in the zone of perpetual darkness, or the geology of how the caves were formed.

Speleology is a science for the dedicated. Where else would a scientist crawl through glutinous mud or climb down a roaring waterfall to examine a rock or collect a bug with only a few soggy pages of an allegedly waterproof notebook on which to record his precious observations? Water, mud, cold, and impenetrable darkness combine to make the work of the cave scientist difficult and slow. To venture beyond the entrance zones, the speleologist must be a thoroughly competent caver. A good proportion of the world's cave scientists are men who first ventured into caves purely for the sport, then becoming hardened cavers, but only later being fascinated by the problems and mysteries of cave science. Speleology is to this day a science with more than its share of mysteries and unknowns, for the nature of the caves ensures that they can hardly ever be studied in the fullest detail. Every time a group or an expedition sets out to visit a cave or region of caves, there is a sprinkling of scientists among those there purely for the adventure and sport of caving; and each cave scientist has his appointed task in the underground world.

Once a new cave has been found, the first priority is to survey it—a slow and laborious task. Maps of caves are very rarely made or needed just for route-finding underground; no matter how complex a cave is, an experienced caver will not normally need a map to find his way through it. But both the intricacies of a maze cave and the incessant twists of a single stream cave make it impossible to the speleologist to maintain his sense of location with reference to the outside world. A cave map is essential so that the hydrologist can appreciate where the water flows underground and how it relates to surface features, and the cave geologist can only really understand the patterns of development in a complex cave with a map of the whole system. A line must be surveyed down each cave passage, using a compass, a clinometer to measure vertical angles, and a fibre tape perhaps 30 metres (100 feet) long, with pocket notebook in which to record the readings. Maintaining accuracy, and observing and sketching enough detail to be able later to draw a useful map, are where the skill and experience of the cave surveyor begin to show. Only rarely can more refined techniques be used.

When the main part of the Pierre St Martin

Cave was discovered in the French Pyrenees there was a plan to utilize the cave river for hydroelectric power by drilling a tunnel in from the valley outside. A very high-grade map was needed, and theodolites were taken into the cave to survey it to engineering specifications. Though it was an arduous task, it was possible in the massive chambers and galleries of the Pierre St Martin. But in a tiny twisting cave passage half full of mud and water, only the crudest and simplest methods work and there is nothing to replace the combination of compass, tape and experience. Dozens of visits may be needed to survey a long and difficult cave, each one involving hours spent lying flat in the mud or crouching neck-deep in cold water trying to make out compass readings in the inadequate light from a caving lamp.

The darkness of the underground is just one more difficulty when surveying caves or studying their geology, but it is actually an advantage in one particular aspect of cave work, photography. The real problems of cave photography are the water and mud and the rough handling which a camera receives on a journey through the caves. Not only has the cave photographer to carry his equipment in heavy, watertight, steel boxes into the cave, but once there he has to be able to clean and dry his hands before extracting the camera from the safety of its box. Lighting the cave presents manifold problems. Flash guns have to be used, and they are notorious for malfunctioning when wet—they rarely last long in spray-filled waterfall chambers. Cave photographs taken with the flash gun simply mounted on the camera rarely succeed. At worst the flash lights up the cloud of steam rising from the wet clothes of the caver, and at best the entire scene just looks very flat and lifeless with a complete lack of shadow. The lighting, therefore, must be

Above : Sleets Gill Cave in England, which consists of a well-rounded tunnel, is a fine example of a phreatic tube

Above : Water pounding down the Wet Shaft in Lost Johns Cave in the English Pennines leaves clean and polished walls that have been cut through the successive layers of limestome
Above right : A narrow passage formed along a joint in the Gaping Gill Cave system, also in the English Pennines, has a floor of boulders which were deposited when floodwaters previously invaded the cave

they flooded or free-draining, that is, phreatic or vadose? The purpose of distinguishing the vadose or phreatic origins of the caves, especially of the old fossil passages, is to relate them to the geological structure and to the topography of the surface at the time when they were formed. In the Yorkshire Dales of England distinct levels can be recognized in the cave systems, above which the old passages are vadose and below which they are phreatic. These levels can therefore be interpreted as the levels of the springs and resurgence caves in the ancient valley floors, before the area was dissected by ice age glaciers. At present caves provide, in this manner, almost the only evidence concerning the glacial re-shaping of the area, and this is fundamental to the understanding of modern landscapes and physical environments.

Observation and interpretation are the classical methods of the geomorphologist both in the outside world and in caves. There are in addition, however, more refined techniques. Cave waters can be chemically analyzed to determine the processes and rates of the limestone solution, and very accurate micrometer-gauges measure the amount of rock eroded from the cave floor. The sediments in the caves can also yield a mass of information. Common in caves are layers of sand, gravel and mud alternating with layers of stalagmite flowstone. Such a sequence means that the cave stream sometimes deposited the sand and mud physically, while at other times it precipitated chemically the calcite flowstone; such changes in a stream's actions can only sensibly be related to bygone climatic changes. In Europe at least they almost invariably relate to the successively different climates of the ice ages. Furthermore the detailed analysis of the tiny amounts of uranium in the stalagmites can indicate, by measuring how much radioactive decay has taken place, the age of the stalagmites. Such studies are currently in progress, in the Yorkshire

remote from the camera, either off to one side or held by a caver some distance in front; the results from the latter method make it known as the silhouette technique. This is where the absolute darkness of the cave is a great advantage. Synchronization and trailing wires can be dispensed with, for the photographer can leave the shutter open while his assistants fire the flashbulbs. He can even leave the camera on a tripod and release a series of flashes himself. The possibilities are many, and this compensates for the hard work and irrefutable element of luck which is involved in all cave photography.

The cave has been found, surveyed and recorded on film: now the specialized cave scientists can move in. The geomorphologist's first task is observation; noting, recording and interpreting passage shapes and details in order to work out how the cave system was formed. Which passages were developed first, and were

Dales for example, which means dating the stalagmites and then relating them to the climatic sequences and also to the morphology of the caves and in turn to the development of the valleys. The caves will eventually do much to provide a complete understanding of the glacial history of the Dales.

All this of course means that the geomorphologist must explore the caves. Personal observation is unbeatable, and even the passages at the very end of long, deep, wet, cold cave systems must be visited to be studied. In such cases it may be an arduous enough task just getting in and out of the cave, but the speleologist must have spare energy which he can devote to careful thought and study while underground. The biospeleologist, on the other hand, does not have to visit such demanding places, for most cave life is to be found in the more accessible parts of the caves.

The cave biologists' difficulties are of a different nature, however. Catching bats which are rabid and infested with malarial parasites is not an enviable task. Bats are not even that easy to catch: success is normally only assured by stretching a mist net, made of very fine nylon, across a cave entrance, and then waiting for them to fly into it. Nets of various sizes are also used to catch the animals in the cave streams and pools, and tiny manually-pumped vacuum-cleaners are used to suck up some of the smallest cave insects. Less subtle techniques are used to catch the larger cave dwellers which crawl about on dry land. The spiders, cockroaches and beetles are not common except in the guano caves of the tropics, and there they are almost too common. In such caves the animals are easy to find and study, but to do so entails walking on and in the guano. It is easy to sink up to the knees in thick piles of bat droppings, yet no footprints remain as it is a living mass of cockroaches and other

scavengers. The guano is thus continually on the move and quickly conceals a footprint.

Fortunately biospeleologists do not always suffer such conditions to study cave life. Cave dwellers can be kept in underground laboratories which are within the cave environment, but which also offer opportunity for study. In the case of non-tropical caves these laboratories can contain far higher numbers of cave animals than are normally found in a single cave.

Many countries have their own cave laboratories, but perhaps the best known are those of France and the United States. The cave of Moulis in the French Pyrenees is a resurgence cave containing about a kilometre (a little over half a mile) of passages, of which the first quarter have now been made into a biological laboratory, with a properly-lit series of workbenches and specimen tanks. The cave air and cave water are used entirely to reproduce natural conditions but in a state where experiments can also be carried out. Some of the Yugoslavian cave salamanders, *Proteus*, are being subjected to a limited amount of light, for example; though still blind they are developing a brown pigment. Some parts of the cave are even less modified. Colonies of various small animals are being allowed and encouraged to thrive in odd corners and under certain boulders where they can regularly be observed. Micro-organisms have even been found in dripwater which is carefully collected from a row of stalactites on the cave ceiling.

The Ozark Underground Laboratory in Missouri is not as old-established or as well-developed as that of Moulis. Its main feature is a roomy stream cave where the water flows over a mass of boulders and stones with a fairly gentle gradient, thus providing an ideal environment for the aqueous cave dwellers. With access by a staircase from the surface, it provides excellent working conditions for the biospeleologists.

Above left : In the cliff shoreline of New Caledonia in the Pacific, the pounding of the waves has enlarged a cave to such an extent that it has cut right through a headland leaving a fine natural arch
Above : Atlantic rollers have etched this cave out of the weaker parts of the rock in the cliffs of the Portuguese coast
Right : The giant natural arch of the Pont d'Arc in the French Ardéche is now just a remnant of a massive cave passage

In terms of numbers, there are probably more cave archaeologists than all the other professional cave scientists put together. In part this reflects the huge proportion of knowledge of Stone Age man which has come from the study of cave homesites. But the archaeologist is only really on the fringe of true speleology, in that he rarely has to venture far underground simply because few Stone Age men did. Not that cave archaeologists do not meet hazards. The Kurnool Bone Caves, in the Erramalais Hills near Madras in India, were excavated late in the last century. Vast numbers of bone artefacts, knives and spears mainly, were discovered, and the great majority were found in the entrance zone. This entrance area, however, is the occasional habitat of panthers and hives of the notorious Indian hornets, both of which can prove fatal.

At Kurnool, and at most other cave sites, the real work of archaeology is the digging and sifting of soft sediments within the daylight zone. This is not always the case, however: in the Cheddar Caves of the English Mendip Hills, a well-preserved skeleton of an ancient inhabitant was found in a sitting position completely buried beneath a layer of stalagmite; in the nearby Wookey Hole, too, many of the best archaeological discoveries have been made by cave divers in the bed of the subterranean River Axe which runs through the once inhabited chambers of the cave; even more remote are archaeological remains preserved in cave chambers, the entrances of which have since collapsed. Such is the case at Stoke Lane Cave, again in the Mendip Hills, where an ancient firehearth has been found in a chamber now only reached by a long crawl down a narrow stream cave and then by a dive underwater into the main caverns.

Many of the world's cave paintings are also found considerable distances into cave systems— our ancestral artists certainly kept their studios well away from their homesites at the cave entrances. The principal difficulty for the archaeologist who studies the paintings is not one of access, however, but of contamination. It is tragic that the fabulous paintings of the famous Lascaux cave are being steadily overgrown by an algae forming because of the humidity increase brought about by the influx of visitors to the cave. Consequently Lascaux has been closed, for there does not seem to be a way in which the original paintings can both exist and be seen.

Geomorphology, biology, archaeology are all relatively 'pure' sciences as far as caves are concerned, with few applications in the practical and commercial world. Certainly geomorphologists have in some parts of the world contributed to the studies of the collapse of cave roofs and subsequent damage, including loss of buildings and roads. But cave collapse is notoriously unpredictable. Other than this, the major practical work of value done in caves has been by the

hydrologists. Cavernous limestone does, of course, contain vast amounts of groundwater, but as a water supply it is rarely utilized. This is partly due to the complex patterns of water flow in the caves. A borehole put down into a good porous sandstone will always produce water once it reaches the water table—the water just flows from the intergranular spaces in the rock. A borehole in a cavernous limestone will only produce water if it hits a cave, and the chances of this are thin indeed. In some parts of the world, however, water is only available from limestone caves, so the problem then is to find exactly where it is and how much there is.

The standard method of determining the paths of underground water is by dye-tracing. Various very powerful, but harmless, dyes can be put into the water which enters caves, and the water is watched for at any of the likely springs or resurgence caves where it returns to daylight. This

Above: The Divaca Cave in Yugoslavia contains spectacular stalagmite decorations. Once a show cave, it has been eclipsed as a tourist attraction by the famous Postojna Cave nearby

Right : The underground River Axe flows through a series of lakes which are the centre of attraction in the famous tourist caves at Wookey Hole in the English Mendips

Right : In the final pool of the Golding River Cave, Jamaica, the water appears a bright green because of the suspended sediment which it contains

method locates the two ends of the subterranean drainage system; in addition careful analysis of all the data on flow-through times, dye-dilution factors, and effects of flooding, can give a fair indication of both the nature of the passages and reservoir capacity of the intervening cave system. In recent years hydrologists have developed tracing techniques utilizing microscopic coloured plant spores and colourless tracers detectable only by their fluorescence, but the older methods of tracing using brilliant green fluorescein dye or bright red rhodamine are still used. Just a few kilogrammes (or pounds) of these dyes in the dry powdered form are capable of colouring vast volumes of water—and indeed enormous quantities are needed to trace large rivers which may take even two or three months to travel through the underground sections. Ideally a minimum amount of dye is used so that the colour is detectable at the springs but is otherwise as unobtrusive as possible. Spectacular results are usually only obtained when the original estimates of flow paths were wrong. A trace with fluorescein on the Hector's River in Jamaica was more than successful. The dye was added where the river goes underground, and it reappeared in the Coffee River resurgence cave 4 kilometres (3 miles) away. The dye was strong enough to colour the entire length of the Coffee River to where it disappeared into the Wallingford Cave 6 kilometres (4 miles) downstream. The bright green

water then emerged from the Mexico Cave another 2 kilometres (1 mile) away and the colour was only diluted to invisibility another 10 kilometres (6 miles) downstream.

Fluorescein was also used to trace a most unusual cave system on the Greek island of Kefallonia. On the western coast of the island a number of fissures on the rocky limestone shore engulf a steady flow of sea-water. Indeed, the flow is adequate to turn the mill-wheel of the Argostolion mill, the only one in the world powered by water flowing out of the sea. For centuries the destination of this water was a mystery, until tracing with fluorescein proved that it re-emerged from springs on the other side of the island, and just above sea level. When the intervening cave system is visualized as a giant U-tube, this uphill flow is, however, easily explained; it is due to the relatively heavy sea-water flowing into the sinkholes balancing a higher level of the lighter, brackish water which emerges from the springs, after it has mixed with fresh water percolating down into the limestone mountains which form the backbone of the island. In many of the Mediterranean countries, notably Yugoslavia, Greece, France, and the Lebanon, limestone hydrologists have been studying the great springs which emerge below sea level just offshore. It is frustrating to see such large flows of good fresh water being lost into the sea because they pass through deep cave systems formed when

Above : The Mas d'Azil Cave in the French Pyrenees carries a small river right through a steep limestone hill. Engineers faced with planning a motor route built the road on a ledge running the length of the spacious cave

70

the sea level was much lower and, therefore, now completely flooded.

One submarine spring on the French coast has, however, recently been developed, and this has involved close co-operation between the engineers, speleologists and cave explorers. Not far from Marseilles, the Port-Miou resurgence produces a large supply of fresh water from a cave 10 metres (30 feet) below sea level. Cave divers explored over a kilometre (half a mile) of completely flooded passages back into the limestone, and realized that a drillhole could pump a fresh water supply from the flooded caves. If much water was pumped out, however, salt water would be drawn back along the caves, contaminating the supply. Consequently the flows of salt and fresh water had to be strictly controlled. This was eventually done when the engineers designed an underwater dam to be built by the cave divers. It was erected in the main passage in such a way as to prevent the heavy salt water from flowing in, while excess quantities of the lighter fresh water could flow out over it. The success of the scheme may provoke the development of other cave rivers in areas where fresh water is at a premium.

Port-Miou, then, was the scene of an engineering achievement in a cave. Its success was founded on the work of the speleologists who studied the cave, but their work depended on

Right : The main gallery in France's Berger Cave is a high tunnel floored with great banks of boulders and clay deposited when the stream runs in flood
Below : A huge chamber in the Palmito Cave, Mexico, has a floor which includes a mountain of broken limestone blocks which have fallen from the roof over many thousands of years

Right: In Israel's Negev desert region, a limestone cave provides water at a spring—a cool contrast to the heat of the surrounding countryside

those who initially explored the cave. Cave exploration is regarded by many as being a sport or recreation, but equally it can be seen as a fundamental branch of speleology, or cave science. Very rarely are caves discovered or explored as part of a professional project. The caves of Kopili in northern Assam, India, are almost unique in this respect: engineers building a dam on the Kopili River had to investigate thoroughly the hydrology of the limestones of the region, and in so doing explored some shallow cave systems containing up to a kilometre (half a mile) of quite large stalactite-decorated passages. Much more commonly though, it is cavers exploring caves for their own enjoyment who produce, almost by chance, dividends for the scientists and engineers.

Nearly all the caves which have been studied by geologists in order to determine the origin and processes of cave development have first been discovered, explored and surveyed by cavers who have done so purely for the recreation. Most of the world's show caves were found by cavers, before being taken over by the commercial developers. Cave explorers, too, have had their share of archaeological discoveries. The clay statues in the Montespan Cave in the French Pyrenees are works of art still regarded as outstanding in the world's richest region of cave art, yet they were discovered by a lone caver who was there because he enjoyed exploring caves. So cave exploration has its place in the science of speleology. A cave research expedition always has its share of members who are there just to find and explore caves. The cavers work very closely with the cave scientist, and many a caver is also a scientist—but even if this is the case a caver will always claim that he goes caving because he enjoys it, because there is something rather unusual and fascinating about caving and being a cave explorer.

Caves around the World

There are caves in Canada and China, in Norway and New Zealand—almost every country in the world has its own caves. Within each country or region the caves have characteristics which distinguish them from caves elsewhere. Although certain features reappear, they cannot just be grouped as 'black holes in the ground'. A journey around the world of caves reveals a variety and range of contrast which reflect the worldwide variation of surface landscapes.

Gypsum caves. Although the great majority of the world's caves are formed in limestone, there are a number of other rocks which may contain caves of quite considerable dimensions. The two most important are gypsum and lava, both of which can support caves rivalling the size of those in limestone.

Gypsum is hydrated calcium sulphate which is highly soluble in water, so there is little difficulty in seeing how caves have developed in this type of rock. In fact, gypsum is at least ten times more soluble in water than limestone and its solubility does not depend on the presence of another substance in the way that limestone requires carbon dioxide. This means, however, that there is less chance for the repeated deposition and resolution which characterizes the reaction of limestone with water—as a result, stalactites and stalagmites are virtually unknown in gypsum caves. In addition, gypsum is much less common than limestone, and few countries have extensive areas of gypsum karst with significant caves. The Harz Mountains of northern Germany are among the best known gypsum regions; dozens of caves have been discovered, most of which are dry, near-horizontal tunnels, but some are of reasonable length and contain large chambers and stretches of active streamway. A much more extensive area of gypsum lies near Sivas in northern Turkey, but it is not so well explored— some caves are known and perhaps more await

discovery. But most amazing of all the world's gypsum karsts are those in Russia. The gypsum of western Russia contains a number of extremely complex maze caves, the total passage lengths of which are remarkable—one of them is the third longest cave system known in the world.

Lava caves. The formation of lava caves does not involve water. These caves are formed when the rock itself forms. Hot molten lava, flowing out of a volcano, cools as it flows farther away from its source. Cooling is greatest at the surface and there is therefore a tendency for a solid crust to develop, while the inner parts are still hot and mobile. Should the inner part than drain out, a cave—a lava tube—is left behind, and within these tubes a continued flow of lava can form a variety of different features. The lava can melt its way into the floor forming vadose canyons, and lava falls can occur where two tubes meet at different levels. Most lava caves are quite close to the surface (their entrances are usually just thin roof collapses), but they can be very extensive. The western United States, Hawaii, Korea, Kyushu Island in Japan, Kenya, the Canary Isles and Iceland are the most important regions containing lava caves. The tube-like tunnels range in size up to 30 metres (100 feet) in diameter, and networks of tunnels may be many kilometres long, but otherwise lava tubes tend to show little variety. Most are gently graded, but if formed in an inclined lava flow they can be quite deep. The Cueva del Viento, on Tenerife Island, has only a short impenetrable boulder pile midway along its length which stops it being followed over its whole vertical range of 478 metres (1568 feet); but nowhere is it more than a few metres below the surface in its course down the long hillside.

The lava caves of the western United States are rather more interesting because of their variety. Long and broad tubes exist in many

Left : Stalactites and flowstone encircle a small pool in Jackson's Bay Cave, Jamaica

75

Above: The ice appears a bright blue when daylight shines down a shaft into the cave beneath the Nigardsbreen Glacier in Norway

states, and the deep tube of Ape Cave in Washington State is only one of a number opened as tourist caves. In the same state, Dynamited Cave is remarkable for its development on different levels, with shafts and solidified lava falls connecting the gently inclined tubes. Even more unusual is South Grotto in Idaho. This is in part of the huge Great Rift—a natural fissure cutting through the Snake River lava field—which has been explored to a depth of about 150 metres (500 feet) before fallen rock stopped progress.

Sandstone caves. These are rarely more than shallow hollows scooped out of cliffs by water or wind erosion, but in some parts of the world, notably in south-western America and central Australia, they have provided useful homesites and are now adorned with fine cave paintings.

Landslip caves. Where natural rifts open up in hillsides when great slabs of rock slip gently downhill, landslip caves are formed. The delightfully named Windypits in north-east England are examples of this, and some fissures are more than 75 metres (250 feet) deep, though half filled with dangerously wedged boulders and rocks.

Sea caves. Rather more varied are sea caves, or littoral caves as they should be termed. Formed by wave erosion they can occur in almost any cliffline or on a geologically ancient or modern shoreline of any sea or lake, and nearly every country in the world has its own examples. Of completely different origin are the great submarine caves known as Blue Holes around the Bahamas Islands. These were formed by normal solution in limestone when they were on dry land during the periods of much lower sea levels in the ice ages; but since then, melting of the great ice sheets has permanently flooded these caves. Explorations by divers, notably around Andros Island, have revealed fine potholes and cave passages including some magnificent stalagmite caverns, all of which are completely filled with sea-water.

Glacier caves. Ice can barely be described as a rock, yet it contains some of the world's most beautiful caves. During the summer season almost all glaciers yield a quantity of meltwater which flows over them, till it sinks into crevasses, and from the bases of the crevasses it carves tunnels through the ice until it can return to daylight at the glacier snout. The movement of the ice means that such glacier caves tend to be rather short-lived, so, although they probably occur all round the world, few are well known. Some exceptions to this occur in the snouts of the much frequented glaciers in the Alps, Norway and the American Rockies. Perhaps the most extensive of all the known glacier caves are the Paradise Ice Caves on Mount Rainier in the Rockies. Wide, arched tunnels, dimly lit by a diffuse blue glow that passes through the thinner parts of the ice roof, are the main passage type,

Above : The cliffs of Aruba Island in the Caribbean Sea contain a wide sea cave where crashing waves demonstrate their erosive powers
Right : The Owyhee River Cave in the north-west United States is a superb lava tube which provides easy walking for cave explorers

and they are adorned with ice stalactites, snow drifts and in some places meltwater rivers.

Limestone caves. Limestones contain the great majority of the world's caves, and even though all have formed in the same rock type there are remarkable variations in their character and morphology, and indeed this is what makes them so unpredictable and interesting. Only occasionally, the nature or type of limestone has a major influence. The caves in the soft porous chalk of northern France are very distinctive, and the porous limestones of the Nullarbor Plain in Australia contain many caves with lakes formed at a water-table due to the porosity of the rock. Most caves, however, are formed in massive fissured limestone, which may appear to be almost identical in many parts of the world.

On the other hand, the structure of the limestone has a considerable influence on the formation of caves. On a small scale, the cross section and pattern of the individual passages are influenced, and on a large scale the whole profile of the cave may be determined. Horizontal limestones contain deep caves which gain their depth in a staircase of vertical shafts, while steadily descending caves generally follow the slope of inclined limestones. Extensive, low-lying plateaus of nearly horizontal limestone are the most capable of bearing long cave systems—the geology of the western Appalachian regions in the United States, which includes the longest

Above: During the Ice Ages, glaciers scraped the limestone clean in the Yorkshire Dales region of the English Pennines

Top: The Vercors region of southern France has not been so recently glaciated and therefore has a thick soil layer and an abundance of vegetation

by several tall, narrow vadose canyon passages.

The combination of high mountain ranges in a tropical region can yield cave development on a spectacular scale, and this is the great attraction of Papua, New Guinea, probably the most exciting area in the world for cavers because of its unique combination of thick limestone, high mountains and tropical climate. Unfortunately, the same factors also make exploration of the island's caves very arduous.

Climate exercises an even greater influence on the calcite decorations in caves—the stalactites and stalagmites—as their formation depends on the movement of carbon dioxide in the water. The whole cycle of carbon dioxide is directly related to surface vegetation, and, in turn, climate. Once the effects of ancient climatic changes have been ruled out, there is a quite clear, general relationship between surface vegetation, landforms and cave decoration. The effect of a moderate climate is represented by the temperate karst of lowland France and Yugoslavia. The rolling surface, pitted with conical depressions called dolines, has a covering of grassland or thin woods, and the caves underground are generally well decorated with calcite.

The warm extreme is represented by the tropical regions, Mexico or Jamaica for example. In these regions the surface is a chaos of limestone towers and deep cockpit depressions, covered by thick jungle. The caves contain huge quantities of calcite deposits; the stalactites and stalagmites grow on a grand scale, almost as thickly as the jungle above. A complete contrast is found in the glacial karst regions. Bare ice-scraped limestone pavements are the characteristic landform, and those around Ingleborough in the English Pennines are fine examples. The cave decorations, however, are not so lavish. The Pennine caves have calcite formations, though it is noticeable how many of the stalactites are just thin frail straws, but the even colder climates over the Norwegian caves account for their almost complete lack of decoration.

One other factor influencing calcite decorations in caves is the geology. A watertight layer of rock sitting above the limestone, for example, would prevent percolation water working its way down into the caves. This is not a common situation, but it can account for some stalactite-free caves. The geology influences the formation of gypsum decorations in a similar way. Being a form of calcium sulphate, gypsum can only occur in a limestone cave where there is an external source of sulphur. One region renowned for the gypsum crystals and flowers which decorate its caves so richly is the eastern United States. Here the main cave-bearing limestone is overlaid by a porous sandstone bed which contains a small amount of pyrite (iron sulphide). As this rusts in the percolating water, the sulphur is carried down to react with the limestone and form the gypsum,

cave system of all (Flint-Mammoth) as well as a number of other really long ones, is of this type. In contrast, very deep caves can only be formed where the mountains are high enough and the valleys are low enough. As a result, Alpine Europe contains most of the world's deepest caves. Steeply inclined, free-draining masses of limestone also encourage deep cave development, and the world's five deepest caves follow the slope of the rock.

Climate has an undeniable influence on cave development. The warm and wet climates of the tropical regions are associated with huge river caves. Jamaica is typical in that large passage sizes and spectacular underground rivers are characteristic of its caves, and this is, at least in part, due to the high rainfall and regular flooding of the caves. An Alpine environment, such as that in Austria, provides the opposite extreme: the bare rock landscapes cannot support surface drainage, thus small cave streams are more common than rivers. More important still, the landscapes of the Alpine regions only date back to the ice ages when they were drastically modified by deep glaciation. The final retreat of the ice left the bare mountains of limestone cut by deep valleys, so the underground drainage rapidly cut down toward the valley floors. While many of the caves have not, therefore, had time to develop into enormous galleries, they tend to include a large proportion of deep shafts linked

which is so abundant in Mammoth Cave, for example.

Clearly it is climate which controls the distribution of the ice which decorates so many caves. Ice caves are those formed in limestone but decorated with ice, as opposed to the glacier caves which are formed in ice. They occur in any cave regions with sufficiently cold climates. The Alps and Pyrenees, Norway and the American Rockies are the principal areas, and in the latter region ice decorates a host of caves formed in both limestone and lava. The atmospheric conditions within the cave which permit ice to accumulate are highly complex. Air is heavier when cold than when warm, with the result that cold air sinks into the caves more readily. Because of this, winter ventilation of the caves is efficient while in summer air movement may almost cease. Consequently permanent cave ice may be established in areas too warm to support the formation of ice on the surface.

Frozen lakes and subterranean glaciers are part of the world of cave ice, but so are the endless stalactites and stalagmites formed in pure translucent ice. Decorated in this way, ice caves tend to be incredibly beautiful, and two very fine examples in Austria have been developed as magnificent show caves—the Dachstein and Eisriesenwelt Caves. Ice crystals are far more delicate; they are very frail and melt easily just from the bodyheat of passing visitors, so it is fortunate that one of the finest ice crystal caves in the world is quite remote. High in the Canadian Rockies, Plateau Mountain Ice Cave has only a few hundred metres of passages, but they are almost completely lined with the most beautiful hand-sized crystals of clear ice—a unique and unforgettable cave.

Geography, climate, topography and geology all account, therefore, for variation in types of caves. Although there is some pattern to the distribution of the different types, one of the most distinctive features of caves is their unpredictability and consequently there is always an element of surprise.

North America
The Green River meanders through the rolling countryside of Kentucky. Its water is sluggish, its gently sloping banks are thickly covered with trees. Almost hidden in the woods on the left bank is a wide, dry cave entrance, and above it the greenery continues to the horizon. Though at first sight it is not an outstanding landscape, it hides well the fact that it is one of the world's most remarkable geographical features. For the cave entrance is that of Mammoth Cave, which extends through a network of 290 kilometres (180 miles) of passages, and is the longest known cave in the world. The low wooded plateaus of Mammoth Cave Ridge and Flint Ridge, just to the north across Houchins Valley, are underlaid by this one huge cave system. A dozen entrances give access to its great wide corridors, vast domed chambers and winding stream passages. The system has very little depth, with few shafts, so the dry main corridors make ideal show caves; many of those in Mammoth Cave Ridge are open to tourists and some of the scheduled tours take up to four hours to walk through the limestone honeycomb. Under Flint Ridge the caves are even more remote and complex, and many are still being discovered and explored.

Long, near-horizontal cave systems, containing both great, dry phreatic tunnels, as in Flint-Mammoth, and meandering active streamways, are typical of the United States' greatest cave region in the Appalachians, stretching from Kentucky up to Pennsylvania and south to Alabama. In contrast, and completely out of character with the region, Ellison's Cave passes

Walls as well as ceiling in Canada's Plateau Mountain Ice Cave are lined with a thick layer of sparkling ice crystals (above), many of which are perfectly formed hexagonal plates (above left)
Top right: Gypsum flowers in Cumberland Cavern in the United States are formed of clusters of pure white curved crystals splaying out from their point of growth on the cave wall
Right: Massive icicles coalesce with a subterranean glacier in the Dachstein Cave in Austria

right through Pigeon Mountain in northern Georgia. It includes ten kilometres (six miles) of passages with entrances at each end, but both entrances lead to immense shafts into the main levels; the two shafts, Fantastic Pit and Incredible Pit, are each about 150 metres (500 feet) deep. Eastern American caves are noted for their lack of extensive calcite decorations. Luray Caverns, a spectacularly decorated show cave in Virginia, and some sections of Mammoth Cave, however, are delightful exceptions to the rule. Virginia is also well known for its Natural Bridge and Natural Tunnel. The former is a massive bridge of limestone linking the sides of a gorge 30 metres (100 feet) wide some 45 metres (150 feet) above its floor—probably the remnant of a once longer cave. Natural Tunnel is less well known but even more spectacular. It is of equally massive proportions: its average height is 25 metres (80 feet), its width 40 metres (130 feet), and it is nearly 300

metres (1000 feet) long. Between its two yawning entrances it is so level and gently curving that a branch line of the Southern Railway has been laid right through it to save tunnelling under the ridge in which it lies.

West of the Mississippi, caves are spread thinly across the United States, but they include a tremendous range of underground scenery. The Ozark region contains the greatest concentration of caves, with such contrasts as the incredibly complex joint-controlled maze of Cameron Cave and the single streamway in Carroll Cave, which is six kilometres (four miles) long. Farther south, Texas boasts the Sonora Cavern, a remarkable show cave with myriads of helictites and crystals covering the walls and the more massive stalagmite formations. The great show cave in the north is Wind Cave, in the Black Hills of Dakota. Altogether it contains 40 kilometres (25 miles) of passage, and its neighbour, Jewel Cave, is twice as long; both are boxwork patterns of tunnels renowned for the layer of sharp calcite crystals which cover much of their walls.

Once into the Rocky Mountain belt, the famous Carlsbad Caverns dominate the scene. The gaping entrance, high in the semi-desert of New Mexico's Guadalupe Mountains, barely hints at what lies below. First is the great Bat Chamber, the roost of millions of bats whose evening flight from the cave is a spectacle in itself. Next is the enormous Main Corridor, strewn with huge boulders, descending steeply to the King's and Queen's Chambers, which contain every conceivable type of calcite decoration. Beyond the chambers is the climax of the cave—the Big Room. It is barely a single chamber, more a massive corridor, curving and broken by pillars of rock; but it is close to 100 metres (300 feet) high, twice as wide and well over a kilometre (over half a mile) long. With its splendid, massive stalagmites and other formations, this has become one of the world's finest show caves; even a fairly rushed tourist visit takes three hours, with a break at the underground dining room, returning to the surface via the 250 metre (800 feet) deep elevator shaft.

North-west of Carlsbad is the Colorado Plateau with its mass of well-known sandstone caves and Indian homesites. Close to Salt Lake City, are Timpanagos Cave, a show cave with a magnificent collection of helictites, and Neffs Canyon Cave, the only one in the United States deeper than Carlsbad. Neffs, however, is a muddy uninspiring cave with little to commend it. In the extreme north-west the Cascade Ranges are famous for their great concentration of lava caves, some of which are very long and deep. An important recent discovery is the Klamath Mountains of northern California; here quite thin, inclined, bands of limestone have in the last few years been found to contain some deep caves. Meatgrinder Cave, for example, is 219 metres (719

Above: The great canyon passage in Fitton Cave, Arkansas, is typical of many of the spacious dry cave systems in the lowland karst of the United States
Right: Delicate glassy helictites coating the stalactites are a feature of the Sonora Caverns in Texas

feet) deep, and consists of rather constricted, steeply descending streamways broken midway by a series of vertical shafts totalling over 100 metres (330 feet) in depth.

Canada was regarded as a country without any significant caves until recent explorations revealed some remarkable cave systems in the very core of the Rocky Mountains. Yorkshire Pot is one of the finest, lying not far north of the border with the United States, with its passages crossing beneath the continental divide. From the bare limestone benches of the Andy Good Plateau a series of vertical shafts leads down 300 metres (1000 feet) to an extensive network of gently inclined phreatic tubes; these are quite large and the name of the main passage—the Roller Coaster Run—reflects both the shape and pleasant nature of the cave. The only cave in Canada deeper than the Yorkshire is Arctomys Pot, high on the slopes of Mount Robson, where a single small stream passage cascades down an endless succession of small waterfalls to a depth of 522 metres (1712 feet).

The most unusual cave in Canada, and indeed unique in the world, is Castleguard Cave. Its single main passage extends for nine kilometres (nearly six miles) in an almost straight line to pass clean under Mount Castleguard, finishing where the passage is blocked with ice underneath the centre of the huge Columbia Icefield. Farther west, in Glacier Park, Nakimu Cave was once a show cave even though it is several hours' walk from the nearest road, the Trans-Canada Highway. The water pounds down a series of cascades in the spectacular river passage and is utterly unapproachable in high summer flow. A few caves have also been found near the Nahanni River in the North West Territories, and though none is very long, they contain some beautiful ice decorations—notably crystals of ice which cover the walls and ceiling.

Above right : Giant Dome in the famed Carlsbad Caverns in the United States is an enormous tiered stalagmite covered with stalactites hanging from its many ledges
Right : A cross section of Carlsbad Caverns shows the inclined Main Corridor leading down to the Big Room, the world's largest underground cavern

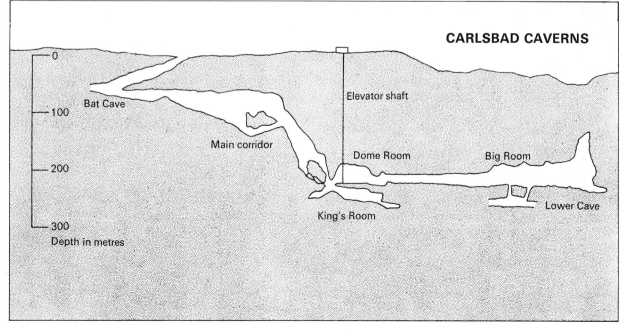

CARLSBAD CAVERNS

Elevator shaft
Bat Cave
Main corridor
Dome Room
Big Room
King's Room
Lower Cave
Depth in metres
0
100
200
300

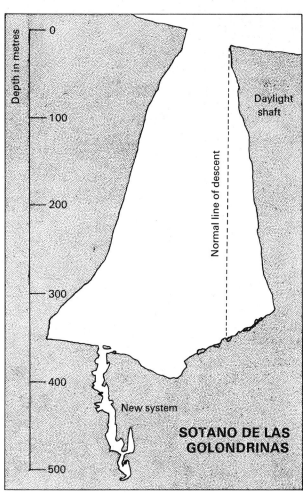

Above : A small river cascades from the entrance of El Chorreadero Cave in the limestone mountains of Mexico
Above right : In cross section, Sotano de las Golondrinas, Mexico, appears as an enormous bell-shaped shaft open to the sky

Mexico

American cavers tend to head south, for Mexico is one of the most exciting cave regions in the world. High mountains, tropical climates, huge rivers and plenty of limestone combine to make an underground paradise. Most remarkable are the huge pits of Sotano de las Golondrinas and El Sotano. With isolated sheer walled shafts 100 metres or so (300 feet) in diameter, the former is 376 metres (1234 feet) deep, and the latter a phenomenal 410 metres (1345 feet) in a single drop. Mexico's eastern mountains also contain the deepest cave on the American continent—Sotano de San Agustin, a classic system of fine stream passages and a staircase of deep waterfall shafts leading down to a sump 612 metres (2009 feet) below the entrance. Fortunately, not all Mexico's caves are deep or wet, and there are also plenty of magnificent great, dry, horizontal systems. A few have been developed as show caves, mostly for their spectacular grandeur, and best known is the Gruta del Palmito, Bustamente, with its mass of stalagmite decoration. The state of Chiapas, in the more remote south of the country, is only just being explored by cavers. It is a land of water caves with huge rivers disappearing underground, some of which have two or three long subterranean sections on their traverse of the limestone country. Among them is the Cueva de El Chorreadero, probably one of the finest in the world. From sinkhole to resurgence it is three kilometres (two miles) long and nearly 350 metres (1130 feet) deep, the whole passage being traversable via a series of waterfalls, lakes, chambers and streamways cut in clean white limestone.

Not so much is known about the caves in the rest of Central America. The Yucatan peninsula of Mexico is practically all limestone, but being a low-level region its caves are mostly the flooded potholes known as cenotes. Guatemala borders Mexico's Chiapas state, so at least in its northern regions the caves are on an almost comparable scale. Of the pits and river caves explored to date, the finest is the Cueva de Agua Escondida, with three kilometres (two miles) of passage, rarely less than 20 metres (60 feet) high and wide, containing a single powerful river so large that even quite a small waterfall is a serious obstacle in the exploration.

The Antilles

Many of the Caribbean islands are also rich in limestone. Barbados and the Bahamas both contain many small caves, but none is long or deep owing to the low relief of the islands. Puerto Rico contains a number of fine river caves, but they are eclipsed by the long and magnificently decorated caves which abound in Cuba; unfortunately, little is known about them.

The island which has attracted most cavers' attention is Jamaica. Much of its surface is

Above: The journey down the Sleeping Pools in Golding River Cave, Jamaica, can be most conveniently undertaken using an airbed as a makeshift boat

formed of cavernous limestone and the upland parts have been carved into vegetation-covered karst; the Cockpit Country has such a chaotic topography of deep depressions that most of it is still unexplored. All over the island there are magnificent caves. Jackson's Bay Cave on the south coast lies at sea level and includes long, fabulously decorated passages containing pools of crystal-clear, brackish water. Further inland are a number of very old caves containing huge, dry tunnels inhabited by thousands of bats and crickets. Thatchfield Cave near the north coast has a long main passage averaging 20 metres (60 feet) in diameter, with a floor of bat guano, forests of massive calcite formations and a gaping hole from the surface with creepers and lianas hanging 30 metres (100 feet) down into the cave. The central mountains in the western half of the island support a number of rivers of which all but one sink underground where they flow onto the surrounding limestone. They reappear nearer the coast at huge springs, such as Dornoch Head, a deep blue pool fed in part by Cave River which goes underground 21 kilometres (13 miles) away. Where these rivers sink, a series of breath-taking river caves occur. One of these, the Quashies River Cave, is known all over the world even though it is not very long or deep. It is cut in clean white limestone and its passages average 5 metres (15 feet) in width and over 20 metres (60 feet) in height. But its real spectacle is the

powerful river crashing over a series of waterfalls up to 30 metres (100 feet) deep.

South America

Africa and South America share the fact that much of their land areas are not developed or thoroughly explored, and certainly cave exploration and study is mainly carried out late in the history of a region's development. Consequently, few caves are known in South America, and it may harbour undiscovered caves and potholes to rival those anywhere else in the world.

At present Venezuela can claim to be the most cavernous of the South American countries. The limestone regions of the north-west contain a host of caves including both deep shafts and massive tunnel galleries. The finest system is the Guarataro, 305 metres (1000 feet) deep, with a series of deep shafts leading to a short active stream passage. In the north-east of the country are the famous Guacharo Caves, populated by thousands of oilbirds—though these distinctive cave dwellers are also found elsewhere in South America. Little is known of Venezuela's south-eastern areas of thick jungle. All that has been discovered so far are the huge potholes of the Sarisarinama Plateau—one is 340 metres (1115 feet) deep—but these may just be a foretaste of what awaits exploration.

Brazil's limestone regions form only a small part of this vast country, and they are not in the

Left: In the heart of the English Pennines a small stream, Fell Beck, plunges down the great shaft of Gaping Gill Hole. Each summer the water is temporarily diverted into another cave and a winch is erected to give cavers and visitors an exciting ride into the depths
Right: The lower half of the shaft bells out into the Main Chamber of the Gaping Gill Cave system and is filled with spray from the lashing waterfall during winter floods

mountainous parts. The deepest cave is, therefore, only 190 metres (620 feet) and this is the Ouro Grosso Cave with a pothole entrance leading into the river cave below. However, the Sao Mateus Cave is still not completely explored, yet already it is known to be 13 kilometres (8 miles) long.

The Andes of Peru contain great masses of limestone, again suffering from a general lack of exploration. The only sizable known caves are at an altitude of nearly 4000 metres (13,000 feet) directly inland from Lima. Here, the Sima de Milpo is a descending stream passage with a series of small waterfalls, where exploration stops at a sump pool 407 metres (1335 feet) below the entrance. The water is next seen at almost the same level in the nearby Huagapo Cave through which it flows to meet daylight at its enormous arched entrance.

Britain and Ireland

The British Isles contain a rich variety of caves, though the low relief of the limestone regions ensures that none is deep in worldwide terms. The Ingleborough-Malham region, in and bordering north-west Yorkshire, contains most of Britain's finest caves. Best known is Gaping Gill Hole, an open shaft which engulfs the stream of Fell Beck on the eastern slopes of Ingleborough Hill. It is a single drop of 110 metres (360 feet), and opens onto the roof of Britain's largest underground chamber 150 metres (500 feet) long and 35 metres (110 feet) high and wide. The stream sinks in the gravel and boulder floor of the chamber, but passages open from the chamber walls leading to a network of more than 11 kilometres (7 miles) of caves. Most are dry, rather uninspiring tunnels, though there are some more large chambers, some active stream passages, a magnificent series of shafts leading in from other entrances, and a few well-decorated areas.

Gaping Gill cannot be called a typical York-shire pothole. What makes the region so fine are the many vadose streamway caves containing relatively small passages descending a staircase of wet vertical shafts. Lost Johns Cave is one of the most popular. A short, wide stream passage leads to a series of junctions where roof passages diverge. The water continues down a magnificent series of waterfall shafts and through some short sumps which require diving gear. Consequently the popular routes down are via the roof passages; these are old, abandoned stream routes with series of fine shafts and galleries which are now dry. All routes unite before a spectacular 20 metre (60 feet) deep waterfall shaft and a winding can-yon which leads to a junction with a main stream tunnel. This is the main drain which collects all the sinking streams over a wide area, and it is 140 metres (460 feet) below the peat-covered fells. Upstream are some climbs and shafts leading up into superbly decorated old tunnels, and down-stream there is a kilometre (over half a mile) of easy walking, interrupted only by a long lake of shoulder-deep water, to a terminal sump. Few of Yorkshire's caves can be entered at their normally flooded resurgences, but one of the finest caves in the area is only entered this way. White Scar Cave has a single main streamway over three kilometres (two miles) long. Low wet crawlways lead in from daylight, but they soon give way to a magnificent vadose canyon. Mostly a metre or two (three to six feet) wide and over 10 metres (30 feet) high, the canyon—richly decorated with calcite throughout—contains a number of low cascades and some long deep lakes. The upstream limit is met where a series of deep lakes, almost filling a large phreatic tube, even-tually lead to a sump. A partly mined tunnel now gives access to the downstream section and part of the streamway has been turned into a show cave. There are a few sections of large, old, high-level passages, however, which are accessible

Overleaf: The Water Chamber in Jackson's Bay Cave, Jamaica, has stalagmites towering over a shallow lake

through the roof of the streamway. The largest section is near the downstream end and there are plans to put a new tunnel up into this, so that tourists can be taken into a massive chamber and the galleries which are beautifully decorated with straw stalactites. If this is done, it will make White Scar the best of Britain's show caves.

Farther south in the Pennines, the Peak District of Derbyshire is another important region of caves. The most spectacular are around Castleton, where a number of shafts and sinkholes feed water down to the resurgence system of the Peak-Speedwell Caverns. In Peak Cavern the main stream passage is a superb phreatic tube, and its lowest section has been turned into a show cave with access through its massive arched entrance. The passages in Speedwell Cavern include some large chambers, though the large main stream pounds its way down a very small, geologically young, passage which is a jagged tunnel in complexly eroded and fretted rock.

The Mendip Hills are best known for their popular show caves at Cheddar and Wookey. Gough's Cave in the sidewall of Cheddar Gorge is a dry but well-decorated old phreatic tunnel, while Wookey Hole has less calcite formations but is the more impressive where the River Axe flows through its various domed chambers. Neither cave compares in length and size with those formed on the top of the Mendips where the streams feed into the limestone. One of the sources of Wookey's water is Swildon's Hole, the longest and deepest in the Mendips. It has a fine decorated stream passage nearly 1500 metres (one mile) long which descends with only a few short waterfalls to a depth of 150 metres (500 feet). A number of sumps have to be dived to reach the end, though some can be bypassed by using the even longer maze of high-level passages which connect to the streamway at a number of points. The high-level caves are much older than the streamway, and consist mostly of quite small, rather muddy phreatic tunnels. Swildon's is probably Britain's most visited, non-commercialized cave, and its popularity is, at least in part, due to its great variety of passages and underground routes. It does not, however, contain any large chambers, so a complete contrast is offered by the nearby GB Cavern which carries a stream on its way towards the springs in Cheddar Gorge. A complex of rifts and tubes in the entrance leads rapidly down to the main passage of GB, which is a steeply descending tunnel, so large that it is referred to as the Main Chamber. Around 20 metres (60 feet) or more high and wide, it has a boulder-strewn floor and stalactite-decorated walls, and it is a very easy journey down to a depth of 130 metres (430 feet). Beyond it are a series of small tunnels and then more large chambers which are decorated with pure white calcite formations.

The limestones around the northern rim of

the South Wales coalfield do not contain many caves, but some of them are among the longest in Britain. In the Swansea Valley lies Dan-yr-Ogof, the only show cave in Wales. The tourist section is in dry decorated tunnels which end at a fine cascading river passage. Beyond this point, only cavers can visit the many kilometres of passage. On the opposite side of the valley lies Wales' best known cave—Ogof Ffynnon Ddu—the longest and deepest in Britain. It has three widely spaced entrances which lead into very complex, gently inclined networks of passages; many are dry, some contain small streams, and parts are extremely well decorated. The total length runs to over 38 kilometres (24 miles). Along the southern edge of these complexes runs the main stream passage, a tall, vadose canyon with few waterfalls but many small cascades and dozens of deep potholes in its floor. From where the water pours through the boulder choke at the upper end of the cave it descends just over 300 metres (1000 feet) to its sump pool a few metres from its point of reappearance in daylight.

Much of western and central Ireland is formed of limestone, and in it have been carved a wealth of fine caves. The greatest concentration is in the Lisdoonvarna area of County Clare. None is really deep for they consist almost entirely of meandering vadose canyons with little gradient and few waterfalls. Coolagh River Cave with its

Right : The tall, narrow streamway in White Scar Cave, in the English Pennines, is so well endowed with ledges that it is easier for the explorer to walk along them rather than the floor
Below : GB Cave in the Mendip Hills of England is best known for the whiteness of some of its stalactite and stalagmite decorations

water-washed flood-prone canyons, and the Doolin Cave system whose many kilometres of explorable passage pass completely underneath the Aille River, are two of the most notable. Ireland's northern region of caves straddles the border between Northern Ireland and Eire. On the Ulster side, the Noon's-Arch Cave system has been claimed as the finest through-trip in the country even though it is only passable by divers. Noon's Hole is a fine 85 metre (280 feet) shaft leading to a near horizontal streamway and then a sump. Cave divers can pass through the sump and explore another three kilometres (two miles) of fine stream passages; eventually these lead out with the water to the mouth of the Arch Cave resurgence.

France

If any one region in the world can claim to be the best for caves, it must be the south of France. The limestones of this region contain most of the world's deepest caves, a fair number of long ones, many magnificently decorated caves, some fabulous painted caves, a collection of superb show caves open to the public, and thousands of other caves large and small. Both the geography and the character of the caves tend to divide the region into three parts: the Alpine fringes, the Causses of central France, and the Pyrenees.

Not far inland from the Mediterranean Sea, the Plateau of Vaucluse is best known for the enormous spring at Vaucluse itself where all the drainage emerges to flow into the Rhône near Avignon. The Fontaine de Vaucluse is a round pool, spectacularly backed by high limestone cliffs, from which a sizable river emerges. Its water rises up a massive tube from a depth of at least 100 metres (330 feet), and it is clearly the outlet of a deep, phreatic cave system. The plateau above contains a pair of deep potholes—the Gouffre du Caladaire and the Aven Jean Nouveau—both of which are about 600 metres (2000 feet) deep. Both consist of steeply descending series of spectacular dry shafts, only in their lower reaches containing tiny streams which have, however, been proved to return to daylight at Vaucluse.

Grenoble lies in the heart of a magnificent region of caves. Limestone mountains loom above the outskirts of the city; the Chartreuse lies on the north side, and the Vercors lies to the south. Most famous of the Vercors caves must be the Gouffre Berger, a cave of magnificent proportions and splendour. Now only third deepest in the world it is still the mecca of the world's cavers as it has only one entrance and a trip to the sump is still the deepest caving trip in the world. A series of shafts and narrow canyons at the entrance make the first 250 metres (800 feet) of the descent very difficult. But at the end lies the huge main gallery. Rarely less than 20 metres (60 feet) wide and 30 metres (100 feet) high—and in many parts

Left: This tall passage in Dan-yr-Ogof Cave, Wales, provides a suitable frame for a small number of very long straw stalactites

much larger—this great tunnel descends gently to a depth of nearly 600 metres (2000 feet); sections of it are superbly decorated with enormous stalagmites and giant staircases of gour pools. This is followed by a section of river passage with long canals and a series of cascades and roaring waterfalls. The last waterfall drops into an enormous sloping chamber, and another series of deep shafts, chambers and long canals finally lead to the terminal sump.

The Vercors also hides a mass of other caves of every shape and size, and fortunately two of the most remarkable are among the most accessible. Opening into the east wall of the Bourne Gorge is the Bournillon Cave with one of the largest entrances of any cave. A tall arch spans the riverbed at a height of over 80 metres (260 feet), and the galleries inside are of comparable size. In the opposite wall of the Bourne Gorge lies the Coufin Cave, now known as the Choranche Cave since it has been developed as a show cave. Thousands of long, delicate straw stalactites are the highlight of the stream galleries and lake chamber which make up this exceptionally beautiful cave.

Of the many deep caves in the sharp limestone mountains of the Chartreuse, the Trou du Glaz system is the most notable because it is also France's longest cave, with over 31 kilometres (19 miles) of passages. The caves virtually honeycomb the great, inclined slab of limestone which rears up to form the Dent de Crolles. There is

Above : A small river tumbles down the boulder-strewn floor of the Metro—part of the main gallery in the huge Pierre St Martin Cave in the French Pyrenees
Right : A beautifully symmetrical group of stalactites reaches into the darkness of the Hall of the Thirteen in France's Berger Cave

one entrance near the summit of the mountain, another at the resurgence down at its foot, and one on each side of the great cliff-ringed ridge. Consequently cavers can enjoy a range of spectacular through-trips in the mountain, passing every variety of active and fossil passage with huge tunnels, narrow canyons and deep shafts.

The rolling countryside of the Causses, fringing the southern edge of the Massif Central, does not have the necessary relief to contain caves matching the depths of those in the Alps and Pyrenees. But some of the impressive caves which occur at lesser depths have been developed as exceptional show caves. The western edge of the region includes the many remarkable painted caves of the Dordogne, and while some are now open to visitors, it is sad that the most outstanding, Lascaux, is closed because of the damage caused to its paintings by the atmospheric changes due to the very presence of the tourists.

Padirac is a completely different type of show cave. A deep circular shaft is descended by a lift, and from the foot of the shaft a huge canyon passage leads to some spectacular lakes and a route which returns through a series of decorated roof galleries. Beyond the magnificent tourist section, the Padirac Cave continues for many kilometres though gaining little in depth; the main problems in its exploration are the many lakes, the high boulder piles over which the

boats have to be carried, and the interminable mud for which the cave is famed. Further to the east, in the Causse Mejean, the Aven Armand consists of an 80 metre (260 feet) deep shaft breaking straight through the roof of a single chamber, which is now also reached by a lift shaft. The attraction in this fine show cave is the 'Virgin Forest', a unique collection of tiered stalagmites which reach heights of up to 30 metres (100 feet) and are almost crammed together in the lower half of the sloping chamber.

The Pyrenees contain an amazing variety of caves. The enormous arched entrance of the Bédeilhac Cave, the show cave on the underground river of Labouiche, the painted caves of Niaux and elsewhere, the immense network of potholes and stream passages which make up the great cave system of the Reseau Trombe and the main road which winds right through the Mas d'Azil Cave are all just part of the eastern Pyrenean underground. Near St Girons is the underground laboratory of the Moulis Cave, which is also renowned for the aragonite crystals in its inner chambers. But the headwaters of the river in the Moulis valley come from an even more remarkable cave—the Cigalère. This is a resurgence cave with passages not far inside it which are unique for the splendour of the gypsum crystals which completely cover their walls. The Cigalère also has a long and spectacular streamway which can only be explored by climbing dozens of cascades and waterfalls.

The western Pyrenees may not match the variety of the eastern area, but they are exciting because of the very deep systems they contain. Exploration within the last few years around the village of Accous has revealed a number of very deep potholes, including two which reach depths of more than 900 metres (2900 feet); the André Touya has a steeply descending passage and then a huge 300 metre (1000 feet) shaft to gain its depth, while the Cambou de Liard has a staircase of more than 50 shafts.

Not far to the west and hugging the frontier with Spain, there is the huge limestone massif which houses a whole series of deep and exciting caves, including the deepest in the world. The limestone mountain slopes down from east to west, and the water which flows to its central part drains down into the underlying cave of Pierre St Martin. It is a slightly unusual cave in that its main passage cannot be entered at either end. In its upstream area a number of large tributary passages join, and then feed into a single river gallery; named after the Speleo Club de Paris which first explored it, this is a fine, tall canyon passage containing long stretches of deep water. Lower down, the river flows through a passage so immense that it is described as a series of chambers: huge boulder-strewn caverns with the river visible only in places as it cascades over and through the boulders for three kilometres (two miles). The last chamber of all is La Verna, a truly enormous void almost spherical in shape and nearly 200 metres (600 feet) across. The river sinks in its floor, but a massive roof passage continues a short way beyond the chamber; from this several systems of narrow canyons and wet shafts reach greater depths, but never rejoin the main river. Five entrances give access to the Pierre St Martin, though none is part of the main passage. Near the upstream limit three series of shafts break into the main galleries, and from the entrance of the highest of these, the SC3 pothole, the total depth of 1332 metres (4370 feet) is measured to the deepest point. A mined tunnel also reaches into La Verna chamber, but the finest of the entrances is the Lepineux shaft. With a single drop of 320 metres (1050 feet) it breaks through the roof of one of the huge chambers in the heart of the system, and much of that chamber is occupied by a pile of fallen rocks nearly 100 metres (300 feet) high.

Italy

The subterranean course of the Reka-Timavo River passes beneath the length of one of Italy's great karst regions, that of Trieste. Italy has many caves, but those of the bare limestone mountains behind Trieste have a special place in history as many were explored in the last century. The deepest in the area is the Trebiciano Cave with a series of mainly short shafts descending steeply to a single massive chamber with the Timavo River flowing across its floor. An even larger chamber is situated only a few kilometres away in the Grotta Gigante. It is at a much lesser depth however, and it is open as a show cave with a steep flight of steps descending a corner of the great stalagmite-decorated cavern—over 125 metres (400 feet) high and long.

The main belt of cavernous limestone forms a

great arc along the southern flanks of the Dolomites and Italian Alps. Right against the border with Yugoslavia lies the high alpine karst of Mount Canin. Over recent years a series of deep shaft caves have been discovered there, and the deepest, the Abisso Michele Gortani, descends via a series of great shafts and tall, narrow meandering canyons to a depth of 920 metres (3018 feet). Farther west the limestones of the Verona area contain the great Spluga della Preta. An impressive great bell-shaped entrance shaft drops 131 metres (430 feet) and leads to a second shaft nearly as deep, followed by a staircase of smaller shafts and narrow intervening canyons to a total depth of 886 metres (2906 feet). The westernmost end of this great limestone arc lies where the Maritime Alps approach the Mediterranean Sea. There the Marguareis Massif is a wild mountanous karst, containing the three entrances to the fine and complex cave system of Piaggia Bella and also entrances to the great shaft caves of Gaché and Cappa among others.

Along the length of Italy, the Apennine Mountains are also rich in limestone and caves. To the north-east of Rome lies what is now Italy's deepest cave. The Grotta di Monte Cucco has a complex entrance series of dry galleries and large boulder-strewn chambers, but farther down a series of deep shafts and sloping tunnels reach a depth of 922 metres (3024 feet) below the upper of the two entrances. Not as deep, but nevertheless a magnificent cave, is the Antro di Corchia in the mountains just north of Pisa. From its upper entrance, a stream passage carrying only a tiny amount of water leads past some large shafts to a series of decorated chambers. Dry passages lead from there to the lower entrance, while the stream runs down a superb passage containing a whole chain of cascades, small shafts and lakes to a huge boulder pile which is the present limit of exploration.

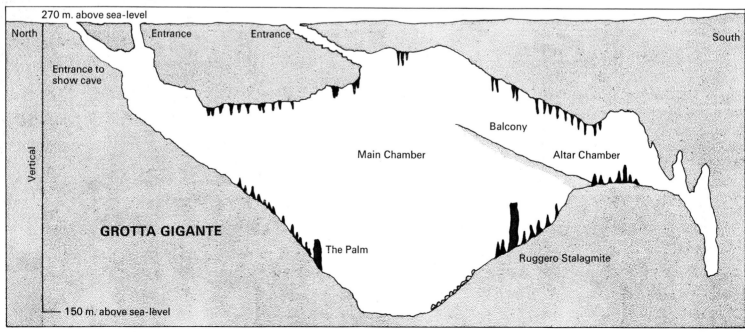

GROTTA GIGANTE

270 m. above sea-level

North

Entrance · Entrance

South

Entrance to show cave

Vertical

Main Chamber

Balcony

Altar Chamber

The Palm

Ruggero Stalagmite

150 m. above sea-level

Left: In the heart of Italy's Antro di Corchia Cave system, a broad ledge takes the caver round a deep shaft
Right: Stalactites and stalagmites completely obscure the limestone wall in the Castellana Cave of southern Italy

Left: In Italy's Grotta Gigante, the great main chamber lies just below the surface, and this cross section indicates just how thin much of its roof is

Spain

Part of the vast Pierre St Martin cave in France loops under the frontier into Spain, and indeed the Lepineux shaft is in that country. But Spain has plenty of other caves, from the spectacular show caves of Drach on Majorca, a number in the Sierra Nevada and the mountains around Madrid to a host in the Cantabrians of the north. The latter is Spain's greatest region of caves, geologically an extension of the Pyrenees. The deepest caves in Spain are found at the eastern end, although few match those in France. Both the Sumidero de Cellagua and the Gouffre Juhué, have entrance systems of shafts leading to extensive near-horizontal galleries with the lowest point in the former being about 1000 metres (3200 feet). Lesser known but nevertheless very fine caves occur along the entire length of the Cantabrians, which include some very long systems, some fine river caves and what is arguably the largest cave chamber in the world in the Torca del Carlista pothole near Santander. But perhaps the best known of the Spanish caves are the shorter ones near the north coast, Altamira and Bustillo for example, which contain some of the finest cave paintings in the world.

Yugoslavia

Half limestone and an important karst region, Yugoslavia contains literally thousands of caves. The few deep ones are tucked away in the Julian Alps in the north of the country, but even better are the well-known caves of Slovenia. For millions of years, the River Piuka has flowed through a low limestone hill just north of the town of Postojna, and in that time it has carved a superb series of caves. The river still occupies the lower levels, but the upper galleries have been left dry and are now decorated with endless varieties of calcite stalactite and stalagmite. Extensive tourist paths and the Postojna underground railway now make many of these galleries accessible for visitors. The Postojna Cave, just above the sinkhole of the River Piuka, is the best known and most visited of the show caves, and contains the best decorated passages. The Planina Cave makes a worthwhile contrast, however, even though it has far fewer visitors; what it lacks in stalagmites it more than compensates for with its spectacular arched river passages.

Besides Postojna and Planina, there are two more equally famous caves of the Slovenian karst. Krizna Cave is not a show cave but is notable for its magnificent passages. Its main tunnel is a tube rarely less than 10 metres (30 feet) in diameter which extends for over three kilometres (two miles). Not only is it superbly decorated with huge stalagmites, but most of its floor is occupied by a series of 22 lakes making it a cave of singular beauty. Skocjan Cave is the last and greatest of these Slovenian caves; its enormous river galleries make it one of the wonders of the world, and developed as a show cave it can be seen properly with the aid of huge floodlights. The River Reka disappears underground at Skocjan, and though it can be followed for nearly two kilometres (just over a mile) in the cave, the water only returns to daylight 33 kilometres (21 miles) away at the Timavo springs in north-eastern Italy.

*Right : Two ice formations—one a jumbled cascade, the other a glass-clear pillar—rest on the solid ice floor of the Casteret Cave high in the Spanish Pyrenees
Below : The cross section through the Antro di Corchia Cave in Italy's Apuan Alps shows how it consists of different series of superimposed passages which lead either to the entrances or the deepest point*

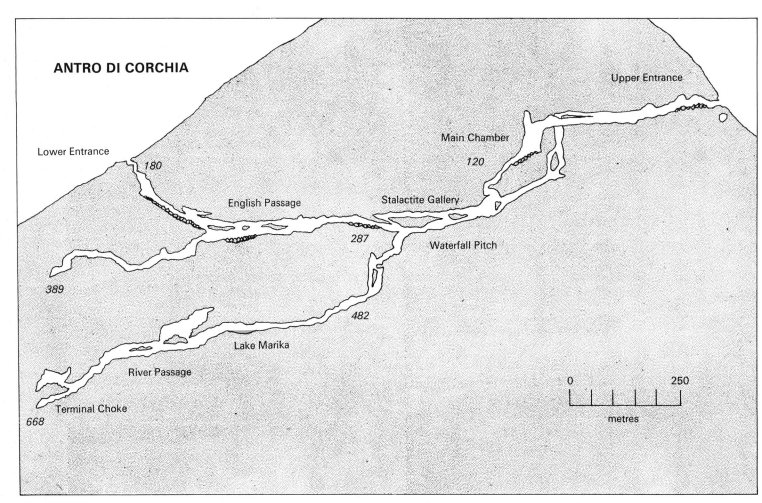

ANTRO DI CORCHIA

Upper Entrance

Lower Entrance

180

Main Chamber

120

English Passage

Stalactite Gallery

287

Waterfall Pitch

389

482

Lake Marika

River Passage

Terminal Choke

668

0   250

metres

Germany and Scandinavia

The northern part of continental Europe has little limestone and few caves. Scandinavia's only important caves are in the mountains of northern Norway, where a number have been explored in thin, sloping bands of limestone and marble. Belgium has only a handful of caves but one of them has been developed as a very fine show cave. The Han Cave has been formed where the River Lesse passes clean through a quite small limestone hill, leaving a series of large and spectacular chambers and a magnificent river gallery down which tourists take boat trips to the wide arch out into daylight. Northern Germany's caves are all in the gypsum beds of the Hanover region. Southern Germany has a number of fine but quite small caves in the limestone escarpment of the Schwabian Alb; in addition, it has just a tiny corner of the Alps tucked into south-eastern Bavaria—and the Alpine regions are rich in caves. The German sector contains the Schellenberg Ice Cave now open as a show cave, and also the country's deepest cave, Kargrabenhöhle, 447 metres (1465 feet) deep.

Switzerland

Although the Alpine mountain chains of southern Europe contain so many of the world's deep caves, the country in the very core of the Alps, Switzerland, contains only a few. Just near

Right : The first explorers of the Planina Cave in Yugoslavia were so impressed by the calcite decorations in the final chamber that they named it 'Paradise'
Below : The hill at Postojna in Yugoslavia is pierced by two long and complex cave systems which carry the River Piuka through the limestone. There is a large region between the ends of the Postojna and Planina Caves that still awaits discovery

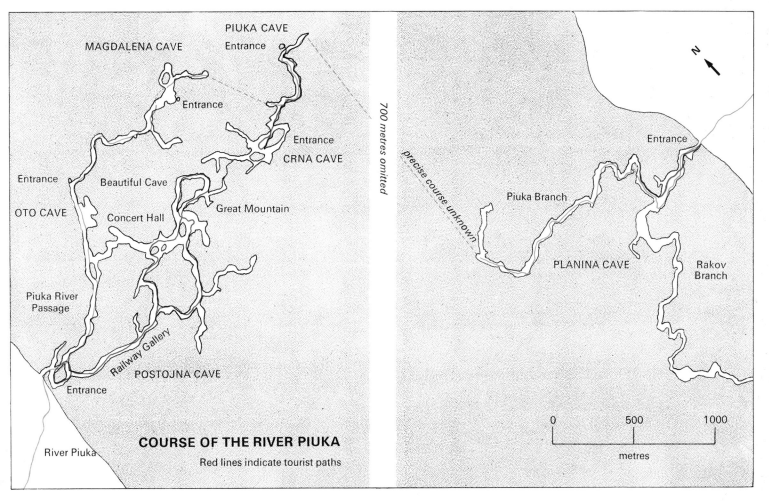

COURSE OF THE RIVER PIUKA
Red lines indicate tourist paths

Lucerne lies the Hölloch, one of the world's longest and deepest caves; it is a huge, inclined network of phreatic tunnels on a grand scale though it has few active streamways or calcite decorations. Exploration is difficult and not yet complete, for it takes a couple of days to reach the highest, remotest galleries which stretch over 600 metres (2000 feet) above the only entrance which lies near the resurgence of the cave's water. Other parts of the high limestone mountains in the heart of the Alps are rather less explored but a handful of deep caves have been found, and there may be more. In contrast, the north-west of the country contains part of the Jura and its generally more extensive shallow caves. The Milandre Cave is a fine example, lying almost on the French frontier; its entrance galleries are dry caverns now turned into a rather unexciting show cave, but beyond the tourist section much better decorated tunnels lead through to a passage containing the large stream which resurges just below the entrance. The stream passages are over six kilometres (four miles) long, and include a magnificent series of cascades, canyons, chambers and decorated galleries.

Austria
At the other end of the Alps, Austria has a tremendous number of caves, not in the high mountains of the Tyrol but in the fringing ranges

Above: Kalvarija in the Yugoslavian cave of Krizna Jama is a magnificently decorated chamber reached only after a long boat ride across a series of underground lakes
Left: Similar in appearance to the pipes of an organ, these stalactites and calcite curtains overlap down the wall of Divaca Cave, Yugoslavia

to the north, mainly in the Salzburg region. Both the Eisriesenwelt and Dachstein Caves are open as show caves, and both are well worth visiting. The show cave sections are richly decorated with ice; massive ice stalagmites and translucent ice curtains adorn underground glaciers which cover the floor of the massive caves cut into the limestone. In each cave the entrance is high up in a mountain face, and at both sites téléphériques have been built to carry tourists to the caves. Many of the undeveloped caves in Austria also contain ice.

The Eiskogelhöhle passes right through a mountain peak, with its vast boulder-strewn galleries decorated with ice for much of their length. In the same region there are immense sub-horizontal systems of massive passages, such as the Tantalhöhle which is over 30 kilometres (19 miles) long, and there are also great shaft caves. Deepest of all is the Gruberhornhöhle which has a series of difficult, wet, shafts reaching a total depth of 854 metres (2802 feet). Even more impressive is the recently discovered Hochlecken-höhle which contains a waterfall shaft 370 metres (1215 feet) deep.

Eastern Europe

Bulgaria and Rumania are similar in that both contain plenty of caves, but none of really outstanding attraction. In each country there are

Right : Massive calcite columns appear to support the wide sloping roof in the Père Nöel Cave in Belgium
Below : Devoid of stalactites, the main passage of Dachstein Mammoth Cave in Austria derives its splendour from the impressive structure of its rock walls

a few caves in the order of 300 metres (1000 feet) deep, and Rumania in particular has its share of long, well-decorated caves. Hungary, too, has caves scattered across its land—none of which is particularly deep—but it does have one famous cave. This is the Baradla-Aggtalek system which, although of no great depth, comprises over 23 kilometres (15 miles) of passages, many containing rivers and numerous calcite decorations. Part of this great cave system extends under the frontier into Czechoslovakia, where it is known as the Domica Cave. Domica has been developed as a show cave, but it is only one of the many fine show caves in Czechoslovakia. In the heart of the Moravian karst lies the gaping Macocha Chasm, an open shaft 138 metres (450 feet) deep which connects with the river passages of the Punkevni Caves which lead to some well-decorated dry chambers. The country shares its finest high karst region in the Tatras Mountains with

Above : Huge ice stalagmites and summer meltwater pools decorate the underground glacier in Dachstein Ice Cave, Austria
Right : In Rumania's Scarisoara Cave the ice stalagmites, each taller than a man, are capped by domes of transparent ice

Poland. The Demanova group of caves contains every variety of decorated passage, river gallery, shaft and ice cave, and various parts have been opened up as show caves. Poland's share of the Tatras is less well explored but does contain one very deep cave; Jaskinia Sniezna is a pothole consisting of cold, wet shafts and narrow passages which lead to a depth of 752 metres (2466 feet) below the higher of the two entrances.

Greece

Greece and Albania both contain huge areas of limestone, though little is known of caves in the latter country. Greece has hundreds of caves, many of which have been brought to light by archaeologists and have revealed rich remains of early civilizations. Some quite extensive well-decorated caves are found right across the country, and some have been developed as fine show caves. Rather less accessible are the deep plunging shafts which are dotted across the Pindus Mountains in the north-west of the country. Epos chasm is the deepest, consisting of a spectacular series of shafts reaching a lake at a depth of 443 metres (1454 feet); and on the same Astraka Plateau lies the single 392 metre (1285 feet) deep shaft of Provetina.

USSR

Not surprisingly the enormous land area of the Soviet Union contains diverse types of karst and caves. East of the Urals the two greatest areas of caves are in the Alaya region on the fringes of the Pamirs, and around Lake Baykal. Perhaps the finest cave in the latter region is the Balaganskaya Cave with its fabulous decorations of ice crystals, but the Alaya region has more caves including some with long river passages and galleries superbly decorated with calcite. In addition, there are the less explored areas. The isolated Balkhan Mountain range just east of the southern Caspian Sea has an arid landscape of bare rock, but the stream courses end in spectacular open sinkholes where they meet limestone. The Urals contain very many caves, in both limestone and gypsum, besides a number of important archaeological cave sites. One of the finer gypsum caves is the Kungur Cave, in the southern Urals, with many kilometres of passages, underground lakes and ice-decorated galleries. Edging the Black Sea, the Crimea and Caucasus contain the majority of Russia's deep limestone caves—including the Snjezhnaja Peshchera, the country's deepest. Exploration of this cave was only halted when a large river was encountered at a depth of 770 metres (2525 feet). Further west still, just north of the Rumanian frontier lies Russia's most remarkable cave region. The rock there is gypsum and in it have formed some of the longest caves in the world. Peschtschera Optimititscheskaja is an incredibly complex maze with a total passage length of 104 kilometres (65 miles).

China

The regions of Kweichow and Kwangsi in southern China contain vast areas of spectacular limestone karst. The tower karst of Kwangsi has often been represented in the pinnacle-like mountains which characterize classical Chinese paintings, and many of the towers are riddled with short caves. Kweichow must boast the more magnificent caves for it contains an extensive limestone plateau, through which many of the major rivers flow underground. Cave exploration is not yet a significant activity in China, but some of the river caves known in Kweichow are of impressive proportions, with passage heights and widths often in excess of 30 metres (100 feet). Paradoxically the best known caves in China are the many small ones in the northern part of the country. These include artificial holes used as houses and cut into the soft loess sediments of the Hwang-ho basin, various caves used as

Above : Tall slender stalagmites form the Bamboo Grove in Demanova Cave, Czechoslovakia

Overleaf : A magnificent series of calcite dams and clear gour pools terrace the wall of the Akiyoshi-do Cave in Japan

Above: Limestone cliffs draped with stalactites make a splendid frame to the entrance of the great Batu Temple Cave near Kuala Lumpur in Malaysia

Island. There the Akiyoshi Plateau conceals a number of caves including one of the same name which has been developed as a show cave. Akiyoshi-do Cave has nearly ten kilometres (six miles) of passages leading to its spectacular resurgence; the show cave is entered at the resurgence and includes a series of magnificently decorated chambers.

India and the Middle East

There are very few caves in peninsular India, though there are a few limestone caves near Madras and some partly artificial sandstone caves at Ellora and Ajanta further to the west. From the little information there is, it appears that the Himalayas are probably the least cavernous of the world's great mountains. Few cavers have looked closely at the region but there have been virtually no reports of cave entrances or karst from the many travellers in the mountains. The only large caves known are quite long, shallow river caves in the foothills of Assam and Nepal. Perhaps the most impressive is in Nepal where the 45 metre (150 feet) deep entrance shaft of the Harpan River Cave engulfs a sizable river just outside the town of Pokhara.

Further west, the extension of the Himalayan ranges contain greater proportions of limestone. The Zagros Mountains of Iran contain only one deep explored cave, but the future should reveal many more. Merging into the cave regions of Europe, Lebanon and Turkey have huge areas of limestone containing long, deep and spectacular caves of all types.

The Sunda Islands

These islands provide both the greatest mystery and the greatest potential for anyone studying caves. Many of the islands contain large areas of thick limestone but the almost total lack of exploration means that little is known of these jungle-shrouded karst regions. On Sarawak virtually nothing is known of the limestone areas in the east of the island, but in Borneo hundreds of caves have been explored. Enormous horizontal river caves are the dominant type, and an example of the dry type is the well-known Niah Great Cave whose vast chambers have for years provided local people with both the salanganes' 'birds-nests' and great quantities of guano fertilizer. To the east, Sulawesi contains areas of spectacular karst but few reports of caves have come from the island as yet. Timor has rather less limestone, but it does boast a river which disappears underground for some kilometres, though cavers have not yet explored the area.

Papua, New Guinea

The high mountain ranges cut by deep gorges are formed largely of limestones, and some of the world's deepest caves may be found on the island. Much is covered by thick, impenetrable rain

temples, and the famous Chou-k'ou-tien Cave fissure which yielded the skull fragments of 'Pekin man'.

South-east Asia

The south-eastern corner of mainland Asia is also rich in limestone. Malaya, Burma, Laos and Vietnam all have sizable caves within their boundaries. Malaya, although probably the least endowed, is the best known for its enormous Batu Caves are religious shrines of major importance. From the other countries, there are only scattered reports of large and long river caves, and if modern techniques of exploration reach these countries they should reveal a mass of superb caves. Japan and Korea both contain extensive systems of lava caves. The Bilremos-gul Cave in Korea is probably the longest in the world formed in lava. Neither country has very much limestone, and the only important karst is on Japan's Honshu

Right : A huge conical calcite boss almost fills the small chamber at the foot of the second shaft of Ghar Parau in Iran's Zagros Mountains

forest. The few travellers bring back endless reports of yawning sinkholes, vast cave entrances, and huge underground rivers. Some of the more accessible river caves have been explored; but the Atea River Cave has proved unexplorable so far because of the enormous and unapproachable river which crashes into it leaving sheer smooth walls in the blackness above. Bibima Cave is the deepest known cave on the island, with a surprisingly easy, gently descending passage finishing in a sump at a depth of 494 metres (1620 feet).

Australia and New Zealand

The vast area of the former contains only a small proportion of limestone karst but, nevertheless, there are some remarkable caves. The more heavily populated south-eastern corner of the continent has seen the most extensive exploration. Hundreds of caves are known there, though none is large on a world scale. The

Jenolan Caves are the best known—developed as show caves, they are renowned for the wealth and splendour of their calcite decorations. Far more significant, though, is the flat desert wasteland of the Nullarbor Plain, which looks little like a karst area. Below ground is a series of fine caves, consisting mainly of huge, horizontal tunnels. Those of Mullamullang Cave are the most extensive and its main passage averages 40–60 metres (150–200 feet) in diameter. Another impressive feature is the clear-blue salt lakes in many of the caves. Weebubbie Cave has a lake 150 metres (500 feet) long with a fine arched roof spanning its width of 30 metres (100 feet).

The rest of Australia contains a real mixture of cave types. There is the Tunnel in the Napier Range of the north-west, a short but magnificently proportioned river cave which is now dry but cuts right through a limestone ridge. In contrast, the caves of the Arnhem Land region, and

indeed much of the Australian interior, are little more than shelters and single chambers, but they are important for their Aboriginal art and burial sites. South-western Australia hides a collection of well-decorated caves. Jewel Cave at Augusta contains some beautiful lake chambers with clouds of straws hanging from the roof. Straws seem to be a feature of this district's caves, and Strong's Cave probably holds the world record with a straw well over 7 metres (25 feet) long. For the sporting caver, Tasmania is the greatest cave region of Australia. Its forest-covered mountains make access difficult and they hide most of the country's longest and deepest caves. Probably the finest system is Khazad Dum. A large stream crashes down a series of shafts leading down to a sump at a depth of 321 metres (1053 feet), and a second cave system carries a tributary stream in through the roof of the massive final chamber.

The two islands of New Zealand each contain their own individual types of caves. On North Island the major group is not far south of Auckland, and long horizontal systems are characteristic. Though there are some fine stream caves and incredibly complex maze systems, the best known cave is Waitomo. Developed as a show cave, the highlight of the tourist visit is a boat ride across a subterranean lake in a chamber, the roof of which is lit by thousands of glow-worms. Completely different are the caves of South Island. Right in the northern tip of the island the alpine karsts of Mount Owen, Mount Arthur and Takaka contain a series of deep and spectacular caves formed in marble. Probably the best example is the deepest: Harwood Hole opens with a huge shaft nearly 200 metres (660 feet) deep, and from the foot of it a passage connects through to Starlight Cave which exits in a gorge 357 metres (1170 feet) below the top entrance. Starlight is a delightful cave as its main passage contains a staircase of giant calcite dams holding back deep gour pools of clear blue-green water.

Africa

Most of Africa's explored caves lie at the northern and southern extremities of the continent. Best known in South Africa are the Cango Caves, whose large chambers, decorated with massive stalactites and stalagmites, have been developed into a very fine show cave. Walkberg Cave contains calcite formations on a similarly grand scale, but the other caves of South Africa are better known for their very delicate formations including complex helictites and superb aragonite crystals. The only deep cave system in South Africa is that of West Driefontein, which consists of a series of vertical fault-guided rift chambers and shafts to a depth of 250 metres (800 feet). Unfortunately exploration has not been completed, and the cave is now being filled with fine

Above: The gaping entrance of the Iaro River Cave looks out into the dense jungle of the southern highlands of Papua, New Guinea

Above: A forest of straw stalactites hangs from the roof of a lake chamber in Augusta Jewel Cave, Australia
Right: Banked flowstone and tiered stalactites cover a wall of the Kubla Khan Cave in Tasmania

mud and tailings from the crushing plant of an adjacent gold mine.

Ethiopia has more than its share of limestone, but the only major cave known is at Sof Omar, 240 kilometres (150 miles) south-east of Addis Ababa, where the River Web goes underground through a spectacular maze of vast tunnels and chambers. Kenya, on the other hand, has almost no limestone and yet it contains dozens of caves, all formed in lava.

Malagasy is a complete contrast to its host continent in that it bears large areas of limestone karst, mainly in the northern part of the island, and in these lie some very extensive river caves.

Northern Africa is different again. As far as caves are concerned, the Sahara is the great unknown. None is likely to be found in most of the totally arid regions of sand, but right in the centre are the caves of Tassili, famed for their ancient paintings of animals who lived there in climates greatly different from that of today. The Atlas Mountains of Algeria and Morocco include considerable areas of limestone with many deep, long caves. The Anou Boussouil in Algeria, one of the finest caves, consists of a fine series of deep, sloping shafts leading to a sump 505 metres (1655 feet) below the gaping entrance arch. Not so deep but equally well known for their large passages are the Friouata and Chikker Caves which carry the drainage out of a closed valley on the northern edge of the Algerian Atlas.

The Lure of Adventure

The underground world of caves offers a unique environment, for it is completely different from anything on the earth's surface. It is also a challenge to visit and so provides fine recreation, together with the mystery of the unknown and the ever-present possibility of discovery.

Caving is a sport which can be enjoyed by practically anyone. Although proficiency only comes with experience, caving is very physical and there are few refined techniques which need to be perfected. Any reasonably fit person with some old clothes, a pair of boots, hard hat and a lamp can go caving. Because of the dangers that await the unwary, by far the best way to start is with an experienced caver, or with a caving club and there is one in nearly every town near a caving area. From the very beginning caving is an adventure so varied that what follows in this chapter can only give a taste of a very unusual sport—cave exploration.

To some people caves offer nothing and they may even appear terrifying. Cavers, on the other hand, are completely happy underground; the darkness does not bother them and they cast barely a thought to the thousands of tons of rock overhead in the relatively tiny cave tunnels. It is only those who feel at home in caves who can continually return to them, either to explore them or just to see them or to work in them. The challenge and beauty of caves, and the thrill of discovery are the attractions, but there are, of course, disadvantages. Caves are always dark, frequently wet, cold and muddy, and are often arduous to explore. But physical discomforts, can be ignored—indeed it is often the uncomfortable caves that are the most exciting. Outside the tropical regions a wet cave is normally a cold one, but usually cavers actually try to follow the water in caves, rather than avoid it. The underground streams have formed the caves and provide the life and activity—in contrast to the

Left : A caver descends a thin wire ladder that disappears into the depths of Eldon Hole in the English Peak District

almost complete silence of the streamless caves. Consequently the most exciting part of cave exploration is in following a stream or river into its underground world deep in a limestone mountain.

Penyghent Pot in the English Pennines is such a stream cave. It is neither the longest, deepest, wettest nor the best decorated cave in Britain, but it is high on the list of almost every British caver's priorities, because it offers a really exciting trip. An unimposing little hole in the peat-covered slopes of Penyghent Hill is the only entrance to the pothole of the same name, and inside the cave a small, rocky and jagged tunnel leads down to a low stream passage. Except in flood conditions, there is only a small stream in the entrance series of Penyghent Pot, but it carries bitterly cold water, and makes the first section of passage quite arduous. For the first 300 metres (1000 feet) the cave is so low that it has to be crawled through, either on hands and knees in the canal section or flat-out in the lowest parts. The end of the crawlway is marked by a noisy cascade, where a wire ladder is used for the descent, which unfortunately has to follow the same line as the waterfall due to the shape of the rock. Below is a passage of another few hundred metres—a stream canyon which is wide enough for comfort but only high enough for a back-breaking, stooping walk. Another short waterfall into a small chamber, and the cave completely changes character.

A low wide passage leads to a lip where the water plunges into darkness 40 metres (130 feet) to the floor of a wide boulder-strewn chamber. The descent is easy by ladders via a series of ledges, but at the bottom the lashing spray of the waterfall, the wind and the noise are altogether awe-inspiring. The chamber and waterfall have been formed because the stream ran into a great vertical fracture in the limestone. The same fracture is followed downstream, in the long, straight Rift Passage which descends steeply as a series of steps. The fracture in the rock acts as a catchment for all the local drainage, and there is a steady fall of water from the roof all along its length, adding to the size of the stream as it crashes noisily over half a dozen waterfalls in the narrow depths of the rift. For part of its length the cavers traverse on ledges in the walls, at other places they climb down in the water. It is an invigorating passage, and the cave seems to beckon onward and downward, following the stream into its mysterious depths.

Then Rift Passage ends and the stream flows into a low, wide passage through a series of pools and down one more waterfall to a deep, wide canal. Just high enough for walking, with waist-deep water, the passage is 5 metres (15 feet) wide and curves off into the distance. The limestone of the walls and roof is nearly black, creating a sombre atmosphere in vivid contrast to the white limestone walls and noise of the Rift Passage. Here the water is silent, and the only sound is the hollow slap of the waves created by the passing cavers. Through a low side passage comes another, even larger stream and the canal ends where the water pounds through a series of pools and cascades. The cave opens out to 5 or more metres (15 feet) square, but the only way forward is in the water, now swift and powerful enough to wash the unwary caver off his feet. There is 500 metres (1600 feet) of this magnificent roaring passage, only interrupted by the aptly named Niagara waterfall half-way along. The noise and sense of power are almost oppressive, and while in its lower reaches Penyghent Pot can be for some, immensely exciting, for others it is a terrifying experience. Then, 1600 metres (one mile) from and 158 metres (527 feet) below the entrance, the river runs into a small sump pool with its only exit under the water. That is the end of the cave for the explorer, except for the return journey which entails fighting the current up the river passage, climbing the endless cascades of the Rift Passage, and finally negotiating the low crawlways near the entrance.

A reasonable time for a small party of cavers to spend on a visit to the sump in Penyghent Pot is about seven hours. And that is seven hours of great sport, hard work and excitement. In that time the caver must walk, crawl, wade, swim, free-climb the rock, and climb swinging wire ladders. Cave exploration rarely demands the technical skills demanded by rock climbing. Caving, however, requires an overall competence that only comes with experience, combined with the ability to accept long periods of time in the discomfort of the water caves. Endurance is needed too, because it is nearly always the farthest reaches or the lowest levels of the caves which are the most exciting, and it is invariably harder to climb out of a cave than it is to descend.

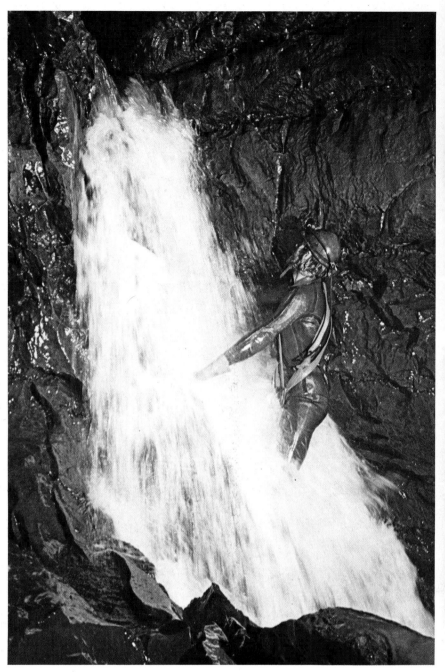

Cold water is a constant feature of caves and caving, and it can sap one's strength more than any other of the natural elements; this is where the high standards of modern equipment are so important.

Today's caver combats the hazard of water with two items of his equipment. Of prime importance is the light, and on explorations lasting no more than a day the electric miner's lamp is unbeatable, with a twin-bulb waterproof headpiece mounted on the helmet and connected by strong cable to the rechargeable accumulator clipped to the belt. The light lasts about 12 hours, and its waterproof properties make it invaluable in some caves. Longer explorations in caves, however, do need lights which can be recharged without an electricity supply: the carbide lamp is, therefore, standard to cavers' equipment. It is fuelled by carbide lumps to which water is added, thereby making acetylene which gives a very

Above: A caver inches his way forward beneath the cascading stream in the lower reaches of Penyghent Pot in Yorkshire
Top right: The most claustrophobic type of cave passage is a long crawlway—such as Hensler's Passage in England's Gaping Gill Cave, where a caver cannot sit up even once during the hour it takes to reach the chambers at the far end

bright light when burned in an open flame—unfortunately, as soon as a waterfall is encountered the light is useless. For wet caves the caver's best apparel is the wet-suit, as worn by skin divers. Made of 6 mm ($\frac{1}{4}$ inch) thick foam neoprene, it is very elastic yet strong, and fits like a glove. It traps water in the bubbles inside it and the lack of water circulation acts as a good insulation jacket, being quite comfortable even in near-freezing water. In dry caves a caver wears any combination of clothing tough enough to stand the hard wear of the jagged rock walls of the caves. A safety helmet and good boots complete the well-equipped caver's outfit.

With wet-suit and electric lamp, water now rarely stops an experienced caver. Sporting caving, or indeed any sort of underground exploration, only advanced beyond the stage of 'peeping into the darkness' about 100 years ago. Late in the nineteenth century caving groups and clubs developed in a number of countries, notably in France, Yugoslavia, Austria, Italy, Britain and the United States. The early explorers, however, rarely went into the wet caves, unless they could take a boat. With no waterproof clothing and only candles in jam-jars for illumination, a deep pool of water, or just a heavy spray from a waterfall, was a serious barrier to exploration. Such conditions prevailed till about the 1930s when the real enthusiasts began to look into the wetter caves, as the accessible dry ones had already been explored. Tweed jackets, cloth caps and candles were little match for spray-lashed waterfall shafts and cascading streamway caves. The cavers fought the elements with strength and will-power, and certainly some of the underground exploits of this era were amazing. In the last two decades, however, modern equipment has been developed, and such obstacles have been almost eliminated. All but two of the world's 24 deepest caves have only been explored

Above: The caver must go slowly through a passage with an airspace as small as this one in England's Peak Cavern, for any waves provide an additional hazard
Right: A cross section through Penyghent Pot, England, shows how its upper and lower series of near-horizontal passages are separated by the waterfalls in the steeply descending Rift Passage

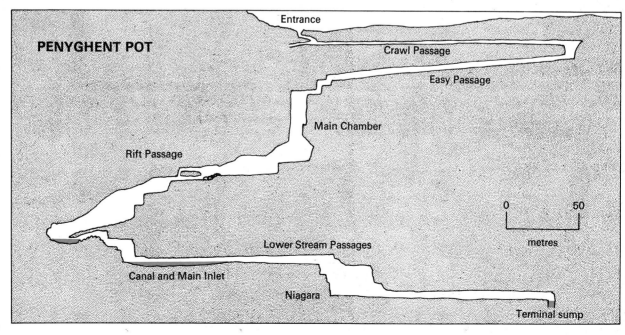

PENYGHENT POT

Entrance

Crawl Passage

Easy Passage

Main Chamber

Rift Passage

Lower Stream Passages

Canal and Main Inlet

Niagara

Terminal sump

0   50

metres

since 1958. Cave exploration today is in the deeper, more arduous caves which were impossible with the equipment of half a century ago. Not only have lighting and clothing been specially developed for caving, but so, too, has the rest of the equipment. Ropes and ladders are essential aids to cave exploration. Straightforward rock climbing techniques are commonly used by cavers, though in the majority of caves they are not adequate. The water-washed, fluted limestone walls of an underground shaft generally offer no holds to a climber. Only laborious 'artificial' techniques, using climbing bolts, would surmount them, so a ladder or a rope offers the most efficient means of descent and ascent. The ladders were originally bulky monstrosities with thick wooden rungs, and the ropes were of none-too-strong hemp; both became fearfully heavy when saturated with water, and 10 metres (30 feet) of ladder or three times that length of

rope was a load enough for any man. But these have been superseded. Electron ladders, made with thin steel wires and light aluminium rungs, weigh no more than a kilogramme per three metres (one pound per four feet), while ropes are now immensely strong, light in weight and usually made of nylon, polypropylene or terylene, none of which absorbs any significant amount of water. Climbing such flexible wire ladders is a skill, and on a long shaft quite a sensational experience, and the ropes are used as lifelines on any but the shortest drop by most cavers.

The most recent developments include the Single Rope Techniques, by which cavers dispense with ladders altogether, slide down the ropes on various friction devices and climb up them using ratchet clamps—a technique which has been adapted from the climbing world. Such methods reduce still further the weights of equipment to be carried down major potholes,

Above left : A stream trickles over a pile of boulders jammed between the walls of white marble in the Norwegian cave of Jordbrugrotten
Above : Much of the length of Canada's Castleguard Cave consists of deep, narrow fissures which can only be traversed by endless climbing—often on wide sloping ledges

Above: The Eroica Shaft in Iran's Ghar Parau Cave provides a fine free-hanging ladder climb

pushed up a shaft like a chimney-sweep's brush; it is then an invigorating experience to climb the swinging ladder hanging from the swaying pole up into a cave passage in the roof. The normal maximum length of a maypole is about 12 metres (40 feet)—longer than that it would be too flexible.

A waterfall too high or smoothwalled to be readily climbed is commonly the end of upward exploration in a cave. Most caves are explored downwards however, and in these cases the most common barrier to exploration is a sump pool—the beginning of a section of completely flooded, phreatic passage. Flooded caves like these are the exclusive realm of the cave diver. Frogmen's diving gear is used, which includes wet-suits, fins, face masks, compressed air cylinders and demand valves to breathe through, plus of course helmets and lamps. The problem is that of carrying the equipment to a diving site at the end of an arduous cave, for the air cylinders in particular are very heavy and must be treated with care. The main hazard in diving stems from the absolute dependence on quite delicate apparatus. Should the apparatus fail, an open-water diver can make an emergency ascent to the surface, but the rock overhead makes this impossible in a cave; cave diving is, therefore, a dangerous sport practised only by a few specialists. It does have its rewards though, such as when extensive dry passages are found beyond quite short sumps. The unpredictability of cave passage patterns means that a diver on an exploration never knows how far he will have to go before finding an air surface at the end of his sump. In many European countries, Australia and the United States dives of over 300 metres (1000 feet) have been recorded. On the other hand some sumps are found to be very short; if these are less than about 10 metres (30 feet) long and not too tight, they can be free-dived with a handline—passed by just taking a deep breath. To free-dive through a sump and then explore the caves on the other side is a memorable experience; the sense of remoteness in a long cave is heightened by the knowledge that the only way back is under the water.

Water gives caving much of its character. Not only does it provide the atmosphere and excitement of Penyghent Pot, for example, but it also supplies the hazards and barriers of waterfalls and sumps. Yet in complete contrast it can offer the most pleasant, easy caving. There is little to compete with the delight of floating on an underground lake in a rubber dinghy. In Yugoslavia, the Krizna Cave is renowned for its lakes, and there are 22 of them in the main passage. Between them are only very short lengths of dry cave, and as the passage averages 10 metres (30 feet) high and wide for a couple of kilometres (over a mile), it is a remarkable cave to visit. The Krizna Cave is not arduous, but it nevertheless attracts

but total dependence on a single rope demands exceptionally thorough care and maintenance of equipment with little room for makeshift adaptation. The success of the Single Rope Technique is well illustrated by the exploration of El Sotano, an immense open pothole in the high karst regions of Mexico, descended by a single free-hanging rope of 410 metres (1345 feet).

While ropes and ladders provide the basic essentials for cave exploration, more specialist techniques sometimes have to be used. Upward exploration of a cave entered at its resurgence is made very difficult where smoothwalled shafts are encountered. Artificial climbing techniques such as drilling holes in the rock and inserting bolts are very laborious and almost impossible underneath a waterfall. For short shafts, at least, maypoles may be used instead. Essentially consisting of joined lengths of scaffolding with a ladder hanging from its end, a maypole can be

In the heart of France's Vercors Mountains, the Coufin cave opens at the base of a great limestone cliff where a small river pours from the rock. Immediately inside the low, wide entrance there is a large lake chamber some 30 metres (100 feet) in diameter, and entering its far side are two spacious stream passages. The lake water is crystal-clear and the streams flow into it through a succession of pools held back by white gour dams. Massive stalagmites flank the walls, themselves covered in stalactites and flowstones. But best of all are the gently arched roofs of both the chamber and the passages, for from them hang tens of thousands of brilliant white straw stalactites, each about two metres (six feet) long. The scene is beautiful, with the various types of formations striking just the right balance.

The Coufin is so accessible and easy to explore that it has recently been made into a show cave, and re-named Grotte de Choranche. Fortunately the development of the tourist facilities has not detracted from the splendour of the natural cave. A caver, however, feels he is almost cheating by being able to see such a fabulous cave with so little effort—there is something especially rewarding about reaching a well-decorated cave after a difficult journey through the underground world.

At the bottom of Strans Gill Pot in the English Pennines lies the main gallery, the Passage of Time. It is spacious with a dry floor and is similar to the Coufin cave in that it, too, is spectacularly decorated with thousands of long calcite straws. Although it is not as magnificent as the Coufin, the Passage of Time in many ways is far more exciting to visit because it is at the end of a difficult cave. Between the gallery and the surface are a series of low, wet passages, four shafts—one of which is 50 metres (160 feet) deep—and a section of viciously tight rifts where cavers have to squeeze between unyielding rock walls no more than 19 centimetres (7½ inches) apart—and few can do it.

The Hall of the Thirteen in the Gouffre Berger is also rewarding. Third deepest cave in the world, lying in the Vercors Mountains of France, the Berger is generally recognized as one of the world's truly fabulous caves. The Hall of the Thirteen is about half-way down the cave, just under 500 metres (1600 feet) below the entrance. To reach it involves a long series of traverses in a narrow passage without a floor, climbing five ladder shafts more than 30 metres (100 feet) long as well as many shorter ones, paddling a boat across an underground lake, and walking and clambering for hours down immense boulder-strewn caverns. Basically a section of the main passage, the Hall is 20 metres (60 feet) wide, and its roof soars to twice that height above a floor which is a network of solid calcite gour dams holding back terraces of deep pools. Almost surrounded by the pools is a magnificent group of stalagmites. Sleek and smooth-sided, up to

Left: The only way back across the churning water in the Quashies River Cave, Jamaica, is by an energetic swim to the end of a ladder Right: One of the more amusing obstacles for a potholer in the Gournier Cave in France is getting out of a dinghy straight onto a swinging wire ladder

experienced cavers for it is a cave explored purely for the pleasures and delights of the underground world.

A very different type of cave, though equally enjoyable, is the Gouffre du Caladaire in southern France, for this one contains no water at all. Instead it consists of a rapid succession of shafts linked by very short sections of passage which are large enough to walk through. The cave descends just over 600 metres (2000 feet) via more than 20 shafts, three of which are over 60 metres (200 feet) deep. It is an enormous task to set up all the equipment needed in the cave, but once this is done travelling up and down the cave is a pleasure. Shaft follows shaft so quickly that each beckons the caver downwards to see what lies below and around the corner. The long climb back up the series of shafts is surprisingly easy as most of the steps in the gigantic underground staircase are quite short, so, instead of being a hard slog, the Caladaire is a pothole to be enjoyed.

Both the Caladaire and Penyghent Pot are notable for the few calcite decorations that they contain. Cavers do not always seek out the decorated caves, for the challenge and adventure of a cave derive from its shape, length and depth, and there is a special type of rather splendid and magnificent beauty in the barest of water-washed caves. Stalactites and stalagmites are not, therefore, the only spectacles of the underground world—they are a bonus.

6 metres (20 feet) tall, they have a symmetry that makes them unforgettable. The Gouffre Berger is cold and wet in parts, so that a full exploration of it is an arduous undertaking, with a few quite difficult waterfall shafts to negotiate. A visit to the terminal sump is, therefore, an exciting challenge, and it is this which attracts so many experienced cavers to the Berger.

As adventure and challenge are two of the main elements of caving, the harder a cave is the better. Such a cave is Castleguard, tucked away in the high limestone karst of the Canadian Rockies. The passages which comprise Castleguard Cave are magnificent; many are geomorphological classics, as is the system as a whole, and they contain some scattered but superb decorations. The most outstanding feature of the cave is the challenge it offers to any who want to explore it to its end—and few have done so. Most of the world's really difficult caves derive some

Left : One of the most pleasant experiences for cavers is to paddle a boat through the long lakes of Yugoslavia's Krizna Cave

Lower left : Long fragile cave straws surround a massive pillar in the Passage of Time in England's Strans Gill Pot

Below right : A section through Castleguard Mountain, Canada, shows how the cave of the same name passes beneath it and out beyond to where it is blocked by the Columbia Icefield

of their quality from their depth. Castleguard, however, goes gently uphill, and the exit journey is downhill. The difficulties of this cave are, firstly, in its remarkable length and, secondly, in the nature of the passages. In addition, the stream system of the cave is fed by underground melt-water from the adjacent Columbia Icefield and may periodically flood and seal off the entrance passages. As meltwater patterns are notoriously unpredictable in mountain regions, exploration of the cave is restricted to winter, when no melt-water exists at that altitude—then freezing conditions are yet another hazard.

The entrance to Castleguard Cave is a wide, inviting archway and the first 75 metres (250 feet) of passage are roomy and horizontal, but end at the lip of an 8 metre (25 feet) deep shaft, at the end of which the cave takes on a completely different character. The wide, flat limestone roof is no more than a metre (three feet) above a most unusual floor of glass-clear ice. In this part of the cave the temperature is still some degrees below freezing, and the ice crawlways are the winter equivalent of a series of canals and lakes which occupy the passage during the summer. The polished ice is very slippery, and, while bags of equipment can be easily pushed along it, cavers have a problem in gaining holds to pull themselves along in the sections which are so low that they can only be passed by crawling flat-out. After 250 metres (800 feet) the ice crawlways end in a comfortable chamber, the centre of which is adorned with three magnificent ice columns.

A few more wide, low chambers which have no ice, lead to another section of low crawl passages. In parts the floor is solid, smooth limestone, but mostly it is a mass of broken boulders and slabs. Progress is awkward and uncomfortable, and it is only the thought of what lies ahead that spurs the cavers on through the miserable tunnel—it is nearly a kilometre

(over half a mile) long with nowhere to stand up in comfort. Worse still, the last of the sections of cave contain three deep pools. As they are so far from the entrance they are not affected by outside conditions and they retain a constant temperature round the year—about a degree above freezing. The pools are thigh-deep and it is impossible to climb round their walls. Walking straight through them is quite feasible, despite the intense cold, but at the start of a long trip into the cave wet clothing is not really advisable, so it is common practise to carry thigh-waders into the cave and don them for the pools section. Once through the pools, the cave enlarges and progress is more rapid with the unpleasantness of the entrance series fading into the distance. For much of the way it is easy walking with occasional climbs over fallen blocks or across huge holes in the floor. The walls are a sombre grey and the floor is dry, brown mud, but a few groups of white straw stalactites act as landmarks on the underground trek. Two kilometres (over one mile) from the entrance the gallery ends at the top of a wide 25 metre (80 feet) deep shaft. The ladder snakes downwards, and at the bottom the cave again completely changes character.

From the base of the shaft, the only way forward is through a tight crawlway, crammed between an arched rock ceiling and a flat floor of hard, dry mud. But after a few metres the roof is higher, the mud ends, and the passage develops into a tubular tunnel, the Subway. Some 4 metres (12 feet) in diameter, its cross section a nearly perfect circle, its walls of faintly banded limestone seem to taper into the distant darkness for it is arrow-straight. Over 300 metres (1000 feet) long and with a gentle uphill gradient, it is like the inside of a gun barrel. It has a weird effect on cavers, seeming almost to beckon them into the black heart of the mountain. The delights of the Subway end where the tube is lost in the roof and

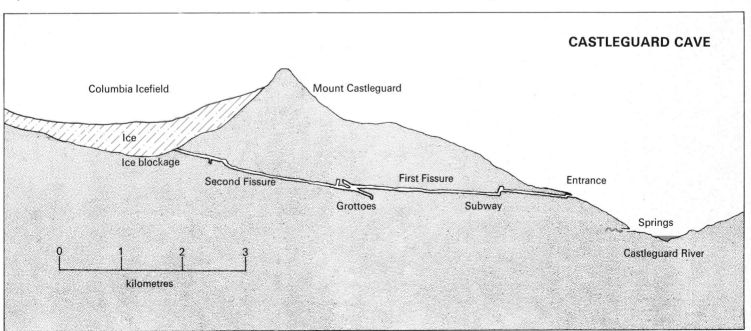

CASTLEGUARD CAVE

Columbia Icefield

Mount Castleguard

Ice

Ice blockage

Second Fissure

First Fissure

Grottoes

Subway

Entrance

Springs

Castleguard River

0   1   2   3

kilometres

Above: Rippled and scalloped ice forms the floor of the series of crawlways near the entrance of Canada's Castleguard Cave
Right: A caver inches his way along a narrow fissure in Castleguard Cave

the passage develops into a tall narrow canyon—and this is where the challenge of Castleguard really starts. The canyon tapers downwards to a slot too narrow to pass along, so the cavers have to climb and traverse the ledges in its walls. Below them the rift drops 10 metres (30 feet) or so; most of it is quite narrow but in parts it bells out into roomy shafts which the caver has to negotiate. The ledges are narrow, covered in mud and sand, and slope steeply into the void. Most of them are only usable when the caver is braced between a pair of ledges on opposing walls, and the constant need to keep in balance to maintain the holds makes the journey an energetic one. Furthermore, the ledges are not continuous, and at many points it is necessary to perform delicate climbing moves to gain more usable ledge systems. For almost its entire length of over one and a half kilometres (one mile) the First Fissure, as it is known, has to be traversed on these ledges as there is little level ground. Heavily laden with a pack full of equipment, a caver may take up to six hours to get through the Fissure.

Following the First Fissure are the Grottoes, a delightful series of passages, which are roomy enough for easy walking—a relaxing contrast to the Fissure. The passages are well decorated with stalactites, straws, helictites, and some beautiful nests of perfectly formed cave pearls. Towards the end of the Grottoes, where the passage has a wide, gently sloping floor of firm, dry mud, is the

campsite. Like almost any underground camp-site when in use it resembles a junkyard, with bits of equipment lying around, muddy clothes from the day's caving scattered over the rocks, and bodies tucking into sleeping bags wherever the floor is level enough. A tiny stream trickles from a hole in the roof and is used as the camp water supply. A thin sweater is all that most cavers have brought by way of dry clothes, consequently the morning awakening—with the search in the blackness for the carbide lamp left ready to be lit, and the single wet muddy boiler suit as the only extra clothing—is one of the less pleasant memories of a long cave exploration.

On the second day of the journey into Castleguard, the last section of the Grottoes passages provides a foretaste. The comfortable flat floor ends overlooking a mud-walled pit with the floor some 10 metres (30 feet) below. Ahead and on the same level the gallery continues, and this is the start of Holes-in-the-Floor Passage. For the next 200 metres (650 feet) the cave has a flat floor of dry mud, except where it is punctuated by 18 holes. These stretch across from wall to wall and are conical pits tapering downwards to where the mud was long ago washed into the deeper parts of the rift passage. They are crossed in various ways: steps can be kicked into the hard, mud walls to traverse most of them, but the one long pit has to be laboriously crossed by chimneying—keeping one's back on one wall and

the feet on the other. Some of the round pits are best negotiated utilizing centrifugal force or the 'wall-of-death' technique which necessitates running round fast enough to reach the other side before falling into the depths. It certainly seems as if Castleguard is trying to protect its innermost secrets, for not far beyond Holes-in-the-Floor the passage again reverts to a narrow floorless canyon, Second Fissure.

Like the First, Second Fissure can only be traversed on sloping ledges connected by delicate climbs. It does have a few chambers with level floors along its length, but for the most part the ledges are narrower, the mud is slicker, the climbs are harder, and the drops below are deeper than in First Fissure—and this time it is over two kilometres (a mile and a half) long. Tortuous in detail and arduous to explore, the Fissure ends at a climb up into a small grotto decorated with beautiful calcite helictites, and from there the passage becomes increasingly spacious and easy. It develops into a tube, like the Subway but elliptical in section, and just where it seems it will continue right through the mountain, it ends in a wall of glass-clear ice which completely blocks the passage. It is an eerie sight—sparkling, inviting, and yet totally impenetrable. In fact it is the bottom of the Columbia Icefield, 300 metres (1000 feet) down from the windswept snows above. The cave has gone right through Mount Castleguard but has failed to reach daylight on the other side.

The spot against the ice barrier is over nine kilometres (nearly six miles) from the only entrance to Castleguard Cave. Because of this exceptional distance from the entrance, it is a true challenge to visit. The only way out for the caver is back the way he came—down Second Fissure, across Holes-in-the-Floor, through the Grottoes, along First Fissure, down the Subway, up the ladders, through the pools and along the ice crawls—another two days back to daylight.

To visit Castleguard's end is a marvellous experience, but this cannot be matched by the unique experience of exploring it for the first time. Like so many caves it was explored in stages, with each successive team pushing farther into the unknown and realizing the excitement of discovery. To tread where no man had been before is tremendously appealing to anyone with a spirit of adventure. Finding caves is a matter of perseverance combined with adequate skills and a great deal of luck—for no one knows where caves are until they are explored. It is the chance of discovery and the excitement that comes with it which is one of the major lures that take cavers to new and distant places.

An expedition went to the Astraka Plateau in Greece during the summer of 1969. Astraka is a wild massif of bare limestone, and the gaping pot-holes which break its surface invited exploration. The expedition, from Britain, was planned with the hope of discovery—the one major pothole had been selected, and would be fully explored. Each member committed himself to the expedition knowing that it could all lead to a great discovery, or a sad disappointment. Early that summer, 15 hardened cavers and 9 expedition helpers, with 24 rucksacks and a dozen donkeys laden with more equipment, set out into the unknown. After a long hot day's march they set up camp on a grassy meadow, not far from a gaping pothole. This was Epos Chasm, found the year before when the expedition leader walked across the Plateau hoping to find caves. He had descended the first two shafts and seen more plunging on downwards, but they were more than he and his few friends could cope with on that first visit. Now he was back and after a night's rest, he led half the team down Epos to see what lay below.

They climbed down the sloping gully at the entrance, and then fed the ladder down the first big shaft. This is a climb of 140 metres (450 feet) to the floor, down a magnificent circular shaft some 10 metres (30 feet) or more in diameter; the ladder hangs free down the centre of the shaft for the last half of its length, providing a magnificent climb as it no longer carries a stream. The floor of the shaft is broken by two open holes. To one side, the larger is an awe-inspiring black gulf, and stones dropped down told the explorers it was at least 150 metres (500 feet) deep. A smaller and more inviting hole opens in another

corner, and so they passed down rolls of ladder and rope, and set off downwards. Next there was a shaft of 45 metres (145 feet) leading to a floor of jammed boulders in a long rift. At one end, another shaft of 20 metres (65 feet) led down to the solid rock floor of the rift. The team climbed down and then walked forward just a few paces until they were on a narrow ledge looking out into the huge shaft they had first seen from above.

The other half of the team took over the following day. Full of anticipation, they set out after a dawn breakfast—who was going to discover the great passages that everyone hoped would lie below? One by one they climbed down the series of shafts, passing down more coils of ladder and rope. Extra ladder was added on in the last shaft and three men descended to a ledge 40 metres (130 feet) below. More ladder was unrolled, one man stayed behind to hold the lifeline, and the other two went down 20 metres

(65 feet) to the next ledge; yet another shaft lay below. Epos was revealing itself as a truly sensational pothole, its downward plunge seemingly without end. One man went down the next shaft, 40 metres (130 feet) this time, and at the bottom he found a passage—was this the beginning of great caverns measureless to man? No. After only a few minutes, the man had found another shaft—a deep one.

The team was now too strung out and further explorations would have to wait for the next day. A pile of equipment was passed down, from man to man, from ledge to ledge, and then began the long haul out. It was decided that evening that one big push would be made the next day with the whole team. If the system continued deeper an underground camp would be made. The next day, all the ledges and shafts were manned again, but this time a group of five stood with the pile of equipment at the previous day's limit. They

fed down a long string of ladder, and, well roped, the first man set off. Some 95 metres (305 feet) below he dipped his foot into the crystal-clear water of a deep lake. The rungs of the excess ladder glinted from beneath its surface. And all around, the sheer, clean limestone walls dropped vertically into the still water: no ledges, no floor, no passages. This remote and beautiful lake, no more than 15 metres (50 feet) across was the end of Epos Chasm—443 metres (1454 feet) down.

So Epos finished a little short of expectation. Geologically it would have been possible for the cave to be nearly twice as deep as it was, but caves are unpredictable and Epos had to be explored to find out what lay below. Although in some ways it was a disappointment that greater caverns were not found, every member of the expedition nevertheless thought the venture worthwhile. It had been a challenge, it had provided excitement and a genuine exploration into the unknown.

Deepest and Longest Caves

The Deepest Caves		metres	feet
Gouffre de la Pierre St Martin	France	1332	4369
Gouffre Jean Bernard	France	1208	3963
Gouffre Berger	France	1141	3742
Gouffre des Aiguilles	France	980	3214
Sumidero de Cellagua	Spain	970	3182
Gouffre André Touya	France	950	3116
Grotta di Monte Cucco	Italy	922	3024
Abisso Michele Gortani	Italy	920	3018
Gouffre du Cambou de Liard	France	908	2978
Spluga della Preta	Italy	886	2906
Réseau Trombe	France	880	2886
Grüberhornhöhle	Austria	854	2802
Hochlecken-Grosshöhle	Austria	819	2687
Hölloch	Switzerland	808	2650
Réseau Ded	France	780	2558
Gouffre Juhué	Spain	775	2542
Abisso Emilio Comici	Italy	774	2539
Snjezhnaja Peshchera	Soviet Union	770	2526
Jaskini Snieznej	Poland	752	2466
Ghar Parau	Iran	751	2463

The Longest Caves		kms	miles
Flint-Mammoth Cave System	United States	290	180
Hölloch	Switzerland	120	75
Peschtschera Optimititscheskaja	Soviet Union	104	65
Jewel Cave	United States	83	52
Ozernaja Peschtschera	Soviet Union	65	41
Greenbrier Cavern	United States	64	40
Reseau de Palomera-Dolencias	Spain	46	29
Eisriesenwelt	Austria	42	26
Wind Cave	United States	40	25
Binkley's Cave System	United States	40	25
Ogof Ffynnon Ddu	Wales	38	24
Sloans Valley Cave	United States	36	22
Cumberland Caverns	United States	35	22
Easegill Caverns	England	33	21
Réseau de la Dent de Crolles	France	31	19
Blue Spring Cave	United States	31	19
Tantalhöhle	Austria	30	19
Réseau Trombe	France	30	19
Carlsbad Caverns	United States	29	18
The Hole	United States	26	16

Glossary

Anastomosing channels: Intricately branching and braided systems of tiny twisting tubes which develop in joints and bedding-planes in limestone as the initial stage in cave formation.

Aragonite: Mineral formed of calcium carbonate, chemically identical to calcite but with a different atomic structure. Under certain conditions it is deposited in caves, usually as sharp-pointed crystals.

Bedding-plane: Break in the original sedimentation of the limestone, which may appear in the rock either as a barely visible line or as an open fracture. Bedding-planes are important in guiding the pattern of cave development.

Boulder choke: Pile of boulders and broken rocks which completely or partially block a cave passage. It may result from collapse or through boulders being washed in from outside.

Calcite: Mineral formed of calcium carbonate which is the major constituent of both cave-bearing limestones and most cave deposits such as stalactites and stalagmites.

Canyon passage: Cave passage formed by the downward cutting of a free-flowing vadose cave stream. Its height tends to be greater than its width.

Carbide lamp: Small headlamp used by cavers and fuelled by calcium carbide and water which react to form acetylene which burns as a bright, naked flame.

Cave: Naturally formed hole in the ground which is large enough to be entered by a man.

Cave pearl: Concretion of calcite, nearly spherical in shape, formed in a shallow cave pool where the water is disturbed, usually by falling drips.

Cave popcorn: Mixed and varied group of cave deposits, most commonly small rounded formations covering walls and surfaces which have been submerged in a pool. Originally an American term, it is probably more descriptive than the synonym 'cave coral'.

Cenote: Pothole partially filled by a lake; named after the many fine examples in the Yucatan Peninsula of Mexico.

Chimney: Narrow, vertical, or near-vertical fissure in a cave which is ascended or descended by using both walls—a method known as chimneying.

Cockpit: Large, irregular closed depression formed in limestone in tropical areas and frequently having a cave opening in its floor. Named after the thousands which form almost the entire surface of the Cockpit Country in Jamaica.

Column: Calcite cave formation which connects floor and ceiling, often resulting from the fusion of a stalactite and a stalagmite.

Crawlway: Cave passage with a ceiling so low that a caver has to crawl.

Curtain: Calcite formation shaped like a hanging sheet, formed by water trickling down a sloping cave ceiling.

Doline: General term of Yugoslav origin for a closed depression, most often conical, in a limestone region.

Dripstone: General term for cave deposits that have been formed by dripping water, and therefore including both stalactites and stalagmites.

Electron ladder: Lightweight flexible ladder made with thin steel wires and tubular aluminium alloy rungs; a standard item of modern caving equipment.

Fault: see Fissure.

Filmwater: Thin film of water which moves slowly over the walls of most active (wet) caves and a significant factor in limestone erosion.

Fissure: Open fracture in limestone formed along a bedding-plane, joint or fault. Bedding-planes occur as a result of sedimentation, and joints and faults are breaks in the rock formed by later earth movements and tectonic deformation. Fissures are an important factor in the early stages of cave development.

Flowstone: General term for calcite cave deposits formed by water flowing over cave walls and floors.

Flute: Sharp-edged groove cut in cave wall by the corrosive action of filmwater.

Free-dive: Dive through a short stretch of completely flooded cave passage without using breathing apparatus.

Glacier cave: Cave formed by meltwater running through or along the base of a glacier.

Gour: Pool of water held back in a cave by a natural barrier of calcite. Water flowing out of the pool over the dam deposits further calcite and thereby builds it up even higher.

Grotto: Loosely used term for a small cave chamber which is attractively decorated with calcite formations.

Guano: Sediment formed of animal droppings. Most cave guano deposits are the product of bats.

Gypsum: Mineral formed of hydrated calcium sulphate which may form complete beds of rock and also delicate crystal formations in limestone caves.

Gypsum cave: Cave formed in gypsum by the solutional action of running water.

Helictite: Small calcite formation, deposited by water, which may grow in any direction, including upwards due to the capillary action of the water in its hollow centre.

Ice cave: Cave in limestone, or any other rock, which contains ice formations throughout the year.

Joint: see Fissure

Karst: Topography and landscape characterized by underground drainage, thus containing sinkholes, caves and springs, with a general lack of surface water and only a thin soil cover. Underground drainage is developed adequately only in soluble rocks, so karst landscapes are found solely on limestone and, to a lesser extent, gypsum. 'Karst' is the German form of the Slovenian word 'kras' which means 'bare, stony ground' and the Kras is the name of the limestone region straddling the Yugoslav–Italian frontier.

Keyhole passage: Cave passage consisting of a phreatic tube with a vadose canyon cut into its floor.

Landslip cave: Cave formed by mass movement of part of a hillside.

Lava cave: Cave formed in a lava flow by solidification of the outer layers while the hot central zone drains away.

Limestone: Sedimentary rock formed by chemical precipitation or the accumulation of animal shell remains normally on the sea floor. The main constituent mineral is calcite which is slightly soluble in rainwater, so caves can be formed in it. Limestone is by far the most important cave-bearing rock.

Littoral cave: Cave formed essentially by wave erosion in a cliff or shoreline of a lake or the sea.

Loess cave: Most commonly an artificial cave hewn out of loess, a soft and easily carved, but adequately strong, variety of silt which occurs in thick beds in some arid regions.

Marble: Variety of rock formed by the action of heat and pressure on limestone. Caves may be formed in it in the same way as in limestone.

Maypole: Segmented scaffolding pole used by cavers for ascending smooth vertical walls.

Maze cave: Complex system of cave passages most commonly formed along closely spaced sets of joints in the limestone.

Meltwater: Water formed largely in summer by the sun warming glacier ice which flows down crevasses to carve out glacier caves.

Network cave: General term for a complex cave system with many intersecting passages.

Percolation water: Water which seeps down slowly through fractures in limestone.

Phreatic cave: Cave formed when completely filled with water flowing under pressure.

Pothole: Vertical-sided cave shaft; also a synonym for a cave system containing a significant proportion of vertical shafts.

Resurgence cave: Cave out of which flows a stream or river; it is therefore explored only in the upstream direction.

Rift: Large or small fissure cave with roughly straight parallel sides, usually formed along a joint or fault in the limestone.

River cave: Cave carrying a sizable flow of water. Exploration requires a considerable amount of swimming or boating.

Rock shelter: Small cave carved naturally into a rock face, usually little more than a simple overhang and rarely containing a dark zone but offering a suitable homesite for cave dwellers.

Sandstone cave: Normally a shallow cave or rock shelter carved into a sandstone cliff by water or wind erosion.

Sea cave: Cave formed by wave action on a shoreline; more strictly known as a littoral cave.

Scallop: Shallow, asymmetrical hollow carved into the wall, floor or roof of a cave passage by eddies in flowing water.

Shale: Sedimentary rock formed of solidified mud and which commonly occurs as thin bands in limestone. It is an important control on the pattern of cave development as it is insoluble.

Single Rope Technique: Method of climbing shafts in a cave by using a single rope and no ladders. Descent is by sliding down the rope using friction devices; ascent is by climbing the rope with the aid of ratchet clamps attached to the body.

Sinkhole: General term for the point where a flow of water disappears into an underground cave.

Stalactite: Calcite formation hanging from the roof of a cave where it is deposited by water dripping out of fissures in the limestone.

Stalagmite: Calcite formation standing on the floor of a cave and deposited by water dripping from the cave roof.

Straw stalactite: Variety of stalactite with thin walls and a hollow centre having the diameter of a drinking straw.

Streamway: Cave passage with a stream flowing along it.

Sump: Section of cave passage which is completely filled with water held back by a section with reverse gradient. It may be passed by cave divers with compressed air apparatus or if short it may be free-dived, but a sump pool frequently bars further exploration.

Traverse: Roughly horizontal climb in a cave, either along the wall of a chamber or using both walls of a rift passage.

Travertine: Hard porous variety of calcite flowstone deposited outside a spring or resurgence cave.

Troglobite: Permanent cave-dwelling animal, adapted specially to underground life.

Troglodyte: General term for cave-dwelling animals, therefore including troglobite, troglophile and trogloxene.

Troglophile: Animal which may live its entire life either in a cave or outside.

Trogloxene: Animal that lives part of its life in a cave but must spend part of its time in the open air, normally for feeding.

Tube passage: Cave passage formed when completely filled with water moving under pressure in phreatic conditions.

Vadose cave: Cave formed by water having gravitational flow and therefore an air surface above it.

Water-table: Irregular surface in the rock, below which all fissures are completely filled with water and above which fissures are only partly filled. It is therefore the boundary between the vadose zone above and the phreatic zone below.

Bibliography

Books on caves range from lightly written descriptions of how to go caving to straightforward guidebooks and academic works on cave science. This selective list includes modern books on exploration and important works on cave science.

AELLEN V. and STRINATI P., *Guide des Grottes d'Europe Occidentale* (Delachaux & Niestlé, Paris 1975)—illustrated guide to Europe's show caves

BAUER E., *The Mysterious World of Caves* (Collins, London 1971)

BEDFORD B. L., *Challenge Underground* (George Allen & Unwin, Hemel Hempstead 1975)—caving in Britain

BERENGUER R., *Prehistoric Man and his Art* (Souvenir Press, London 1975)—painted caves of northern Spain

BÖGLI A. and FRANKE H. W., *Radiant Darkness: the Wonderful World of Caves* (Harrap, London 1967)—photographic book

CADOUX J., *One Thousand Metres Down* (George Allen & Unwin, London 1957)—exploration of the Gouffre Berger, France

CASTERET N., *Ten Years under the Earth* (Dent, London 1963)—exploration in the French Pyrenees

CHEVALIER P., *Subterranean Climbers: Twelve Years in the World's Deepest Cavern* (Faber, London 1951)—exploration of the Trou du Glaz caves, France

CULLINGFORD, C. H. D. (Ed.), *British Caving* (Routledge & Kegan Paul, London 1962)

HALLIDAY, W. R., *American Caves and Caving* (Harper & Row, London 1974)

HEAP D., *Potholing* (Routledge & Kegan Paul, London 1964)—caving in Britain

HERAK M. and STRINGFIELD V. T. (Ed.), *Karst: Important Karst Regions of the Northern Hemisphere* (Elsevier, Barking 1971)—a regional geography

JENNINGS J. N., *Karst* (M.I.T. Press, Cambridge, Mass. 1971)—geography of karst and caves

JUDSON D., *Ghar Parau* (Cassell, London 1973)—exploration of Iran's deepest pothole

LÜBKE A., *The World of Caves* (Weidenfeld & Nicolson, London 1958)—caves and exploration in Austria

McCLURG D. R., *The Amateur's Guide to Caves and Caving* (Stackpole, Harrisburg, Pa. 1973)—caving for beginners in the United States

MOHR C. E. and POULSON T. L., *Life of the Cave* (McGraw-Hill, New York 1967)—photographic book on cave-dwelling animals

ROBINSON D. and GREENBANK A., *Caving and Potholing* (Constable, London 1964)—a book for beginners

SMITH D. I. (Ed.), *Limestones and Caves of the Mendip Hills* (David & Charles, Newton Abbot 1975)—a regional geography

STENUIT R. and JASINSKI M., *Caves and the Marvellous World Beneath Us* (Kaye & Ward, London 1966)—caves and caving in Belgium

SWEETING M., *Karst Landforms* (Macmillan, London 1972)—geography of karst and caves

TRATMAN E. K. (Ed.), *Caves of North-West Clare, Ireland* (David & Charles, Newton Abbot 1969)—a regional geography

VANDEL A., *Biospeleology: Biology of Cavernicolous Animals* (Pergamon Press, Oxford 1965)—textbook on cave biology

WALTHAM A. C. (Ed.), *Limestones and Caves of North West England* (David & Charles, Newton Abbot 1974)—a regional geography

WALTHAM A. C., *Caves* (Macmillan, London 1974)—illustrated general book

Index